SHALE BOOM

OTHER BOOKS BY DIANA HINTON

Author

Morpeth: A Victorian Public Career

Coauthor

Oil Booms: Social Change in Five West Texas Towns

Wildcatters: Texas Independent Oilmen

Life in the Oil Fields

Easy Money: Oil Promoters and Investors in the Jazz Age

Oil and Ideology: The Cultural Creation of the American Petroleum Industry

Oil in Texas: The Gusher Age, 1895–1945

SHALE BOOM

The Barnett Shale Play and Fort Worth

DIANA DAVIDS HINTON

Fort Worth, Texas

LIBRARY OF CONGRESS CATALOGING-IN-PUBLICATION DATA

Names: Hinton, Diana Davids, 1943- author.

Title: Shale boom : the Barnett Shale play and Fort Worth / Diana Davids Hinton.

Description: Fort Worth, Texas : TCU Press, [2018] |

Identifiers: LCCN 2018016244 (print) | LCCN 2018017385 (ebook) | ISBN 9780875656946 () | ISBN 9780875656854 | ISBN 9780875656854?q(alk. paper)

Subjects: LCSH: Gas wells--Texas--Fort Worth Basin--Case studies. | Shale gas reservoirs--Texas--Fort Worth Basin--Case studies. | Gas wells--Texas--Fort Worth Basin--Hydraulic fracturing--Case studies. | Oil wells--Texas--Fort Worth Basin--Hydraulic fracturing--Case studies. | Gas well drilling--Environmental aspects--Texas--Fort Worth--Case studies. | Shale gas industry--Social aspects--Texas--Case studies. | Petroleum industry and trade--Social aspects--Texas--Case studies.

Classification: LCC TN881.T4 (ebook) | LCC TN881.T4 H56 2018 (print) | DDC 333.8/233097645315090511--dc23

LC record available at https://urldefense.proofpoint.com/v2/url?u=https-3A__lccn.loc.gov_2018016244&d=DwIFAg&c=7Q-FWLBTAxn3T_E3HWrzGYJrC4RvUoWDrzTlitGRH_A&r=O2eiy819IcwTGuw-vrBGiVdmhQxMh2yxeggw9qlTUDE&m=N5ffg8lZ-x3f4aoMve_myozo3i6veMI8dtcKgWXviVI&s=iw2Pyuowop4x4DoU1wNv5oI-oL6VrjdJF5YM8zVol4M&e=

TCU Box 298300
Fort Worth, Texas 76129

817.257.7822
WWW.PRS.TCU.EDU

To order books: 1.800.826.8911

TEXT AND COVER DESIGN BY ALLIGATOR TREE GRAPHICS

For
Nina and Mike and
Mel and Glen

CONTENTS

ACKNOWLEDGMENTS

TO BEGIN, I WANT TO THANK THE PERSONS WHOSE MOST generous help and encouragement made it possible to research and write this book.

The project began with a suggestion from Texas Christian University Provost Nowell Donovan that the impact of the Barnett Shale gas-drilling boom on Fort Worth and its surrounding metroplex was worth scholarly study. In the spring of 2008, TCU Press Director Judy Alter and Professor Greg Cantrell asked me if I would be interested in taking on such a study. At that time, the boom was nearing its peak, and I had been following its progress with great interest. The opportunity to study what was happening was irresistible.

When I mentioned the project to my good friend Glen Ely, he told me that one of his neighbors, Marty Searcy, was one of the Four Sevens group, key independent players in broadening the scope of the Barnett action to include the Fort Worth metropolitan area. Marty Searcy gave me an invaluable introductory overview of the play; he also identified some of the most important leaders in the action, thus making it possible for me to contact them. Above all, he helped me understand the economics of doing business during the boom and the problems independents faced when gas drilling entered city limits.

Marty Searcy was the first of many with persons who were willing to give hours of their time to share their experiences with me. Some, like Dan Steward, helped me grasp the geology and technology of the Barnett play. Others, like Don Young, pointed out the environmental problems posed by gas drilling. All have uniquely valuable perspectives to offer, essential to going beyond the simple sequence of events to presenting the human side of the boom. What interviewees shared brought the story of the boom to life.

Those interviewees also include: Judy Alter, David Arrington, Larry Brogdon, Jerry Cash, Ted Collins, Decker Dawson, Nowell Donovan, Melinda Esco, Roger Harmon, Ed Ireland, Steve Jumper, Jack Ladd, Dick Lowe, George Mitchell, Charles Moncrief, Mike Moncrief, Ken Morgan, Hollis Sullivan, Ray Tobias, David Watts, Dan Williams, and George Young Jr.

I also wish to thank people who helped me in many other ways. Linda Bomke, Dick Lowe, and David Arrington were especially helpful in finding other materials for the story. TCU Press came forward with support to launch the study. My friends Nicholas Taylor, Ralph Veatch, and Patrick Whelan, expert respectively in law, petroleum engineering, and geology, gave the manuscript of this book close scrutiny for the sorts of inaccuracies or omissions that can happen when translating petroleum technicalities into terms accessible to those unfamiliar with the industry.

In researching and writing this book, I have been especially fortunate to have had the resources of the J. Conrad Dunagan Chair of Regional and Business History at the University of Texas of the Permian Basin. Chief among the benefits offered by the chair has been the opportunity to gain the help of a brilliant research assistant, Judy Madison. Judy has done outstanding work in tracking down materials essential to telling the story of the boom and its aftermath. Her work kept progress on the project going even when medical problems forced me to take time out. Beyond any doubt, without Judy's dedication, this study could not have stayed on course to completion.

Throughout my work on this book, my late husband Harwood

Hinton, always a patient and cheerful partner, gave me continual loving support and encouragement. I owe him most thanks of all.

Diana Davids Hinton
The University of Texas of the Permian Basin

INTRODUCTION

IN 1956, M. KING HUBBERT, A GEOLOGIST WORKING FOR SHELL Oil Company, predicted that oil production in the United States would peak in the early 1970s and steadily decline thereafter. Coming after years of postwar boom in the oil patch, during which time the petroleum industry found millions of barrels of new reserves, Hubbert's predictions were, to say the least, startling and apparently counter to common sense. But, confounding his critics, Hubbert proved to be right—in terms of what happened during the next half century: US oil production peaked in 1970 and then began what seemed an inevitable decline.[1]

Hubbert's pessimistic outlook was not the first of its kind. Partly because many oil wells come in for relatively strong production only to have production wane over time, even in the American petroleum industry's earliest years, there were those who wondered how long "rock oil" production would hold up. In 1909, David T. Day of the United States Geological Survey (USGS) offered the alarming forecast that US oil production would be insignificant by 1935. Ironically, in that same year, thanks to the bonanza discovery of the giant East Texas oilfield, the American petroleum industry struggled to right itself after a crisis brought about by markets awash with too much oil. Yet even during that crisis, there were prophets of future oil famine who warned that too much oil on the market encouraged waste of a finite resource it had taken millions of years to create.[2]

So Hubbert was not the first industry observer to forecast future oil shortages, nor would he, by any means, be the last. The energy crisis in the 1970s, largely the outcome of an economic witch's brew of increased energy demand, low petroleum prices, misguided federal regulation, and international politics seemed to make it obvious that the prophets of shortage were right. America's production of oil and gas had nowhere to go but down. Even as global petroleum markets suffered from a glut of oil in the later 1980s, the pessimists remained credible. They included highly knowledgeable industry observers like Princeton University geologist Kenneth S. Deffeyes, who, in 2001, predicted that global oil production would reach its maximal level in 2005. In a 2008 edition of his work, *Hubbert's Peak: The Impending World Oil Shortage*, Deffeyes noted that global oil production "stopped growing" in 2005, promoting a crude oil price spike of $140 a barrel. He had said there was "a geological limitation to the oil supply in the ground," and, as he put it, "The urge to say, 'I told you so,' is too much to resist."[3]

And yet as I write this, the global petroleum industry is once again in economic crisis from overproduction, as the supply of oil has greatly outstripped demand. Starting in June 2014, oil prices have dropped from roughly $100 a barrel to levels below $30 a barrel. The United States is now the world's largest combined producer of oil and gas. In 2014, the United States produced an average of 8.7 million barrels of oil a day, outproducing Saudi Arabia, and oil imports fell to a five-year low. During the same year, natural gas production averaged 74.7 billion cubic feet a day, keeping gas prices low and confounding expert opinion of only a decade earlier that the nation would need to import increasing volumes of natural gas to meet demand. The amount of oil and gas in storage has shot up, with oil stockpiles nearing record levels.[4] Once again, the prophets of shortage are wrong. But why? The developments responsible for beginning a dramatic reversal of national petroleum fortunes are the subject of this book.

Before going further, let me stress that the prophets of shortage, for the most part, have not been misguided or ill-informed industry observers. Nor would I argue that a future petroleum shortage could not happen. The history of the petroleum industry has always been full of surprises. Forecasters not only have to work with known sources of petroleum, but they must also base their estimates of future supply on available technology and prevailing prices. You may know the petroleum is there, or suspect it is, but is it technologically possible and affordable to get it out of the ground? When Hubbert made his prediction in 1956, he was working with known reserves exploited by the exploration and production technology of his time—which also happened to be a time of falling petroleum prices and rising industry costs. In effect, for his time, he was right. Similarly, if you were an American geologist in 2000, it was reasonable to think that most of the world's petroleum reserves, accessible and affordable using conventional methods, were already known. What turned dire forecasts upside down was the use of new technology to access oil and gas in shale. This breakthrough came in the Barnett Shale, and it has had immense national and global significance. The prophets of shortage did not foresee that.

To explain how technological changes could have so enormous an impact, it is necessary to introduce some greatly simplified petroleum basics. Oil and gas come from organic matter contained in sedimentary rock. They can move through some types of rock and not others, moving until trapped by folds or faults in rock they cannot penetrate—salt, for example. The rate at which oil and gas move through rock depends on the rock's permeability. They may move quite readily through sandstone but take millions of years to migrate through dense, minimally permeable rock like shale. Regardless of the natural permeability of the rock containing oil and gas, if that rock can be broken or shattered around a wellbore to create cracks or openings through which oil and gas can move easily, it would be possible to get out much more oil and gas in a far shorter time than it would take to produce them from unbroken rock. Production is thus stimulated by what amounts to an artificial enhancement of permeability.

Over time, the petroleum industry has used various ways to open up rock around a wellbore. Early Pennsylvania oilmen discovered they could bring greater production from a new well by pouring nitroglycerine down the hole and detonating it, a technology they dubbed "well shooting." Shooting was a haphazard way of breaking up rock, as well as a dangerous one, since nitroglycerine sometimes detonated unexpectedly in unintended places, like the well shooter's truck. In the 1930s, oilmen found they could stimulate production from wells in limestone formations by pouring acid down the well to dissolve and acidize the rock. But in the late 1940s, the industry developed a far more efficient stimulation technology in the form of hydraulic fracturing, generally called "fracking." This technology consists of injecting fluid under enormous pressure into rock around the wellbore, creating cracks kept open by sand or some other proppant in the fluid.

One challenge fracking posed to petroleum engineers lay in figuring out what sort of fluid to use. Engineers experimented with a wide variety of fluids augmented by various chemical concoctions, but they learned experimenting was not without its risks. Use the wrong fluid, and one might inhibit rather than promote production. They also learned that the fluid that worked well in one area might not work well in another. And they found some types of formations that defied their best experimental efforts—shale, for example. There were other places, like the Permian Basin's Spraberry Trend, where fracking the Spraberry boosted production at first, but after several months, production dismally declined. In other words, as a stimulation technology, fracking was clearly an improvement but no magic bullet.

Production stimulation was not the only challenge oilmen faced when they drilled conventional vertical wells. In looking for oil, what the prospector hoped to encounter was a petroleum-bearing rock formation many feet thick. Drilling down vertically through a thick pay zone meant the wellbore would expose many feet of oil- or gas-bearing rock. But in any oil field there might be places where the petroleum-bearing rock was thick and places where it was not. Drill into a thin spot, and even fracking would not turn that vertical well into a huge producer. Here the technological breakthrough came in the late 1970s with the development of flexible drill pipe and horizontal drilling, permitting a horizontal wellbore. A vertical well through ten feet of pay might not be worthwhile, but what one might get by drilling a mile or so horizontally and staying within a thin pay zone was another matter. Horizontal drilling permitted tapping rock underneath obstacles like towns and highways. It also offered the chance to drill a number of wellbores from one central well pad, especially desirable if the rock targeted—rock like the Barnett Shale—contained a great deal of petroleum locked within a small area. Used with fracking, it could offer a new opportunity to go back to old areas seen as largely played out, as well as try areas hitherto unpromising, like the Fort Worth Basin.

This story of the Barnett Shale boom begins with the region in which it took place. As a petroleum-producing region, by 1990 North Texas seemed over the hill. Its oil fields had never been as large or lucrative as, for example, many in the Permian Basin or on the Texas Gulf Coast. As for the Fort Worth Basin, wildcatters there were more often disappointed than lucky, though deeper drilling in the 1950s brought in substantial reserves of natural gas in some places. Outside the urban centers of Fort Worth and Wichita Falls, North Texas remained predominantly rural. Its economy was dominated by agriculture, but frequent drought and lack of abundant groundwater made agriculture more challenging than lucrative. Its oil industry saw two decades of feverish exploration in the teens and twenties, which transformed Fort Worth and, to a lesser extent, Wichita Falls, into regional oil centers. Thereafter, its oil activity was eclipsed by the East Texas boom during the Depression and sidelined by wartime shortages of material. The postwar years brought a renaissance of exploration, resulting in discovery of natural gas at depths greater than those tapped earlier. But in the 1960s, most major companies shifted prospecting away from North Texas. The major oil companies were hunting for elephants, and North Texas seemed like rabbit country—best left to independents.

If one went looking for independent oilmen, one could find plenty of them in Fort Worth. From the city's beginnings as a cattle town and rail hub, Fort Worth evolved into a major regional petroleum industry center. By the 1920s, it was a home to oil field service and supply companies, pipeline networks, and refineries. Major companies active in

North and West Texas had large regional offices there, and there was also a large and thriving independent oil community. Despite bumpy industry fortunes in the Depression, Fort Worth independents survived and prospered, mainly by exploration and production outside North Texas. In the 1940s and 1950s, the wealthiest men in town included oilmen like Sid Richardson, Amon Carter, and W. A. Moncrief, who had made fortunes in oil fields elsewhere. There were also scores of independents ready to try their luck at projects closer to home that were likely to have outcomes too modest to tempt major companies. After 2000, such men would play a critical role in developing the Barnett Shale. Before that, however, it was an independent outsider from Houston, George P. Mitchell, whose aggressive and tenacious strategy unlocked the enormous reserves of the shale.

Were this study to have a hero, it would be George Mitchell. Mitchell was the pioneer of shale production because he was willing to challenge conventional thinking. Finding possibilities in what others dismissed as impractical was, in fact, a key element in his business strategy. Thus, he entered North Texas action in the 1950s by taking on a project that had been rejected by many another oilman, a project passed along by an investor's bookie. He focused on finding natural gas at a time when gas prices were so low that only a giant find was likely to make much money. To grow, he took on a breathtaking amount of debt. Perhaps most important, he was willing to work with the relatively new technology of fracking. When existing fracking formulas did not produce commercially attractive results in the Barnett Shale, he bullied his engineers to come up with one that did. What they found, using slickwater fracking, was completely at odds with prevailing opinion on accessing shale. In any event, many other oilmen saw Mitchell's obsession with getting production from shale as either misguided or deluded—that is, until he was successful, accessing a staggering amount of natural gas, economically recoverable reserves estimated at fifty-three trillion cubic feet in 2015.[5]

Long before anyone knew how much gas there might be in the Barnett, however, local independent oilmen noticed that by 1998 Mitchell was doing a tremendous amount of exploration and infill drilling within a core area of southeastern Wise and southwestern Denton Counties. The tantalizing question they could not answer was whether the Barnett would be productive much beyond the area Mitchell had under lease. The question was especially difficult to answer because no one outside Mitchell's staff had given the Barnett much study, and Mitchell Energy was not about to share information it had spent millions of dollars to get.

Still, Mitchell's escalated drilling encouraged smaller independents to try their luck. Here the vanguard included partners Dick Lowe, Hunter Enis, and Larry Brogdon; Dallas independent Trevor Rees-Jones; Ted Collins and George Young; and Hollis Sullivan. They began leasing land outside Mitchell's core turf, taking leases on "ranchettes," four- to six-acre properties outside city limits, and moved south into Parker, Tarrant, and Johnson Counties. Many property owners, novices at oil leasing, were willing to lease for relatively low figures, like $100 to $150 an acre, sums affordable for small independents. When these independents combined Mitchell's slickwater fracking technology with horizontal drilling, they began to bring in production substantial enough to get the attention of larger independents like XTO Energy, Devon Energy, and Chesapeake Energy, who proceeded to buy them out. The boom got under way.

Any boom presents a familiar set of problems to communities, problems that may be anticipated but not readily resolved. The Barnett Shale boom was no exception. As rigs appeared in fields and pastures, small towns struggled to meet the needs of a horde of newcomers. There was never enough housing to meet need; roads could not stand up to heavy industrial traffic; employers faced sudden labor shortages; demand for goods and services was round-the-clock. The boom gained momentum as aggressive leasing moved from the semirural environment of ranchettes to suburban neighborhoods. Oblivious to problems a drilling rig down the block might create, home-lot owners began to unite to drive tougher bargains with landmen over the amount of bonuses and royalties, haggling both up to new heights. As homeowners hurried to get on the leasing bandwagon, so did public and private institutions. Cities leased parks and airports; school districts, colleges, and universities leased campuses; churches and clubs leased whatever open land they had: all looking forward to "mailbox money."

When rigs began to move into neighborhoods, however, both city governments and gas drillers faced unfamiliar situations. Cities had to allow drilling—that was a matter of Texas law—but on what terms would it take place? There were virtually no guidelines to direct how far rigs should be from homes, schools, or public institutions. How would cities manage traffic around drill sites or nuisances like round-the-clock noise and lights at rigs? Safety issues had to be considered. Even if gas well blowouts were not likely, they were possible; what would happen in the event of disaster? Gas drillers had no easy answers to these questions, for in some respects they were as ill-prepared for drilling in an urban setting as cities were. Historically, most oil and gas exploration had not taken place in town. Industry players were used to cutting deals with farmers and ranchers, putting down wells in the wide-open spaces. Accustomed to regulation by state agencies like the Texas Railroad Commission (Texas RRC), they were not receptive to following a set of city rules governing rig-site operation. Nor did they expect to encounter any resistance to gas drilling. After all, why would anyone oppose making money?

During 2007, Barnett Shale action accelerated dramatically throughout North Texas, peaking in the summer of 2008. The region experienced tremendous growth in both new jobs and wealth. In the city of Fort Worth, the *Star-Telegram* ran a daily blog tracking the boom, and there was a surge in construction of office space and apartment units. Larger independent companies engaged in fierce competition for drilling sites and drove lease bonuses to unheard of heights. Meanwhile, gas drillers took wellbores under all sorts of unlikely places—the Texas Christian University (TCU) football stadium, downtown's Sundance Square, and the runways of the Dallas/Fort Worth International Airport (DFW). Yet, for all the boom-time excitement, critics of industry operations grew in numbers and visibility. Attempting to head off opposition, large companies launched a variety of pro-drilling public relations projects whose message was clear: everyone benefitted from the boom.

And then boom gave way to bust. Natural gas prices fell, and with them the value of gas drillers' company shares. The leasing bubble burst in October 2008; companies trimmed back lease bonus offers prior to halting leasing. When companies scaled back, thousands of landmen lost jobs. Thereafter, the rising tide of layoffs hit oil field and construction workers. As natural gas prices continued to fall, royalty checks shrank. Cities and institutions began to face budget problems, going from windfall to shortfall. Thus, within a space of six months, the Barnett Shale boom ended. But a national controversy over industry operations and technology essential to the boom was just beginning.

At first, industry opposition had a local focus; as boomtime euphoria passed, neighborhood residents of Fort Worth and surrounding smaller cities began to have second thoughts about drill sites near homes and schools, pipeline construction through front yards, and noisy compressor stations. City councils faced pressure for more comprehensive and aggressive industry regulation, pressure that met heavy-handed industry resistance. But in 2009, regional industry opposition moved to a broader environmental focus in which two issues assumed major importance. One was air pollution. Was natural gas drilling and processing not only responsible for increasing pollution but also for releasing toxic carcinogens in residential areas? The second was the technology key to bringing natural gas out of shale: fracking. Was this technology responsible for natural gas polluting groundwater? Was it even causing earthquakes?

Environmental opposition directed toward air pollution and fracking gained the traction that would move it to national attention outside the Fort Worth Metroplex, among residents of rural areas in Wise, Denton, and Johnson Counties. They were not happy to see industry invade the countryside. When they complained about industry operations to state agencies, in particular the Texas Commission on Environmental Quality (TCEQ) and the Railroad Commission, they received minimal response from regulators. In turn, rural industry opponents hired environmentally friendly consultants to test their air and water and to develop evidence of industry wrongdoing. They used that evidence to press for federal action from the Environmental Protection Agency (EPA). Not only did their strategy resuscitate the old power struggle between state and federal regulatory agencies, but it also brought angry property owners' concerns to the attention of national environmental advocacy groups like Earthworks and the Sierra Club. Barnett activists would find additional allies when fracking technology was transplanted from North Texas to Pennsylvania, where it opened up the vast gas reserves of the Marcellus Shale. That brought fracking to the doorstep of Northeastern environmentalists, and the media began to find opposition to fracking made good copy. Ban fracking, as New York ultimately did, and one banned the technology making gas drilling worthwhile.

Ironically, for North Texas opponents of gas drilling, neither attempts at federal intervention nor pressure for a "moratorium" on gas drilling brought the Barnett Shale boom to a standstill. Instead, market forces in the form of an oversupply of natural gas brought down gas prices and, with them, the boom. But by 2009, when the boom was obviously over, controversy over emissions, and more especially fracking, reached beyond a national to a global audience. At home, the EPA struggled to devise regulations for gas-drilling emissions and to decide whether fracking really polluted groundwater. Abroad, the French simply decided to ban fracking within their national borders. Meanwhile, geologists in many countries pondered whether the new technology could open up bonanzas on their home turf. Environmentalists mobilized to block it.

Beyond question, the Barnett Shale boom has had an enormous impact on the petroleum industry. The technology developed in the Barnett that made the boom possible has opened up vast new oil and gas reserves in the United States in formations like the Marcellus in Pennsylvania, the Bakken in North Dakota, and the Eagle Ford in South Texas. It has permitted oilmen to reassess known producing areas of marginal profitability in the West Texas Permian Basin and tweak the technology to yield a staggering increase in oil production. Reserves now accessible in the Permian Basin's Midland Basin alone have been estimated at seventy-five billion barrels. Such development offers the prospect of many decades of oil and gas dominance of the American energy supply. In effect, the innovations launched in the Barnett Shale have had as great an impact on the petroleum industry as the innovations launched by John D. Rockefeller—and they have been just as controversial.

Though controversy is unavoidable in telling the Barnett story, I have not intended to offer a brief for either the petroleum industry or its critics. Rather, I have tried to present the perspectives of the various stakeholders caught up in the boom. The boom created challenging situations, many without precedent, to all who were part of it. The homeowner signing the first minerals lease; the gas driller sorting out the logistics of a well inside city limits; the city council member trying to decide how best to regulate industry operations in the public interest; the environmental advocate sounding the alarm that drilling threatened both nature and the public—all confronted problems as unfamiliar as they were immediate. Whether in the end the boom benefitted those within it can be debated, and has been. That it made a tremendous difference to a region, a nation, and an industry is undeniable. Opening the Barnett Shale began the process by which the United States may now access billions of barrels of oil and trillions upon trillions of feet of natural gas, petroleum reserves rivaling those of the Middle East. That is why the story of the Barnett Shale is worth telling: its impact has reached beyond its region to the nation and the world.

CHAPTER 1

Oilman's Graveyard?

IN 1990, IF YOU HAD ASKED SOMEONE IN THE PETROLEUM industry where the next Texas boom would be, you might well have been met with a stare of disbelief. On the heels of the great price-crash of 1986, it was hard for anyone in the oil patch to believe there would ever be another boom in Texas. Even if your industry observer could muster up a few shreds of optimism, the last place an oilman would have been likely to suggest for a new boom would have been North Texas or, within it, the Fort Worth Basin. That a boom a dozen years later would unlock trillions of cubic feet of natural gas would have seemed completely impossible. No oilman familiar with regional history would have predicted that.

As a petroleum-producing region, North Texas seemed over the hill by 1990. Having produced millions of barrels of oil for close to a century, the region's aging fields now typically produced from thousands of wells whose production was dwindling to marginal levels—stripper wells maintained by tiny one- or two-man firms struggling to make ends meet in industry downtimes. Within North Texas, the Fort Worth Basin would have seemed even less likely as the source of a petroleum bonanza than other sites. For decades, the basin both tantalized and frustrated most prospectors who tried their luck: it tantalized them because in so many places there were surface signs of oil and gas underground; it frustrated them because these surface indicators were so often misleading, resulting in drilling dry holes. At best, when petroleum exploration paid off, it offered modest returns. With a few exceptions, the Fort Worth Basin's fields were generally small and scattered, not likely to make anyone's fortune when oil prices were lower than twenty dollars a barrel. And in some parts of the basin, Tarrant and Johnson Counties, for example, no one had ever found a commercially viable amount of oil or gas at all. If ever a place could be called an oilman's graveyard, the area around Fort Worth qualified for the title.

Historically, the region containing the Fort Worth Basin was not easy for early settlers to penetrate and develop. What geologists call the Fort Worth Basin is largely coterminous with what geographers and ecologists identify within Texas as the western Cross Timbers and the Grand Prairie. Topographically, the Cross Timbers and Grand Prairie contain rolling hills cut through by creeks and rivers, the largest being the Trinity and Brazos Rivers. Today, outside the Metroplex area, stretches of open grassland mix with extended, once exceptionally dense stands of post oak and blackjack oak, gnarled trees usually growing no more than thirty to forty feet high. Explorers of Texas found the stunted forests unusually challenging to negotiate, often having to axe their way through close-ranked trees and saplings interspersed with briars. The author Washington Irving, who accompanied Henry Ellsworth's expedition into the Oklahoma Cross Timbers in 1832, found the terrain so rugged he thought the woodlands like "a forest of cast iron" after a few hours' ride within it shredded his coat.[1] The Hudson Valley woods were tame by comparison.

Notwithstanding barriers put in place by nature, American settlers gradually entered the Cross Timbers, their migration accelerating after Texas joined the United States in 1845. Many of the immigrants came from the upper South, particularly the states of Tennessee, Arkansas, Missouri, and Kentucky. Most were not rich, and unlike planters farther south on the Texas Coastal Plain, most did not own slaves or raise cotton. Instead, they settled where grassland met forest, making a living by hunting, ranching, and subsistence farming. The grasslands encouraged raising cattle; the timberland hogs, for the hogs fattened readily on the superabundance of acorns in the oak forests. Settlers put land to the plow to raise corn and wheat, but by the time of the Civil War, corn was the essential crop and corn bread the settlers' dietary staple.[2]

North Texas was not an easy place for a farmer to make a living. Most settlers led bleak, hardscrabble lives with few material comforts. Shortly after the Civil War, journalist Stephen Powers, who was walking across the South to the Pacific Coast, described the home of a settler living not far from Johnson County:

> His log-cabin stands inside of the rail-fence circle, and in the dreary yard there is not a bush, absolutely nothing else but the pyramidal ash-hopper standing on its head. In the door sits his sallow wife, barefoot and with disheveled hair, her elbows on her knees and her chin elevated across her hands; while a group of weasel-faced children and a monstrous brindled dog squat about. . . . In the field there is some pale phantom corn . . . sharing the ground with the sumac.

The householder himself was in bed with chills and fever, perhaps malaria.[3] Allowing for Powers's negative view of Texas and Texans, he probably did not exaggerate the grim surroundings he saw.

After the Civil War, cotton joined cattle as a source of agricultural income in North Texas, but the longer-term benefits of cotton cultivation were questionable in the counties whose economies it came to dominate. Not only did short-run profit from cotton discourage crop diversification and encourage landowners to push tenants or sharecroppers to grow cotton, but cotton took a heavy toll on the soil. Grown on sandy soil, it drained away nutrients and left land vulnerable to erosion. Worse yet, if cotton could withstand extremes in weather, it was vulnerable to devastation in the form of the boll weevil, which moved from Mexico into Texas in the 1890s. By World War I, that insect was decimating cotton fields in North Texas, and as it ravaged crops, it did similar damage to the incomes of those who raised them. By 1930, the golden days of cotton in North Texas were but a memory.

Where cotton had once grown there remained thousands of acres of eroded fields and gullies. As Richard Francaviglia has noted, federal land experts cited the Cross Timbers region as "a textbook example of land abuse."[4]

Though the economy of much of North Texas was solidly tied to agriculture, railroad construction in the 1880s and 1890s encouraged a start at developing the region's mineral resources, since railroads offered a cheap way to ship stone, sand, and gravel used in building and road construction. Regional clay deposits encouraged brick manufacture. Deposits of soft coal, albeit laced with sulfur and slate, prompted development of mines at Bridgeport in Wise County, Newcastle (appropriately and ambitiously named!) in Young County, and most notably, Thurber in Palo Pinto County. Since the railroads were dependable consumers of coal until they switched to oil after 1900, the coal companies had a ready market for their output. The resource most needed in a region where drought was common, however, was good-quality drinking water, and to the chagrin of many a farmer and rancher, it was not always easy to find.[5]

From settlement onward, well diggers found plenty of water unfit for consumption. As USGS geologist C. H. Gordon would observe in 1913, in north central Texas, "all underground waters are mineralized to a greater or less extent." Some wells produced water so heavily laced with mineral salts that it was not only undrinkable but too mineralized for boiler use. In 1893, for example, the Fort Worth and Denver City Railroad's attempt to put down a water well at Rhome, in southeastern Wise County, brought in water so laden with iron salts and sulfur it could not be used in locomotive boilers. Fifteen years later, some Rhome businessmen considered putting this sulfury water to constructive use by imitating Mineral Wells to the west and establishing a "health spa"; the plan did not advance beyond discussion stage. In some parts of Tarrant, Denton, and Dallas Counties, wells dug deep enough could tap into drinkable water in the Trinity aquifer. That was the regional exception.[6] Even worse, all too often when well diggers reached water, they also brought up oil and gas.

Throughout North Texas, as in many other parts of the state, small seeps of oil and gas were extremely common. Geologist W. M. Winton, writing on the geology of Denton County in 1912, noted that at the base of the water-bearing Trinity and Paluxy rock formations, there was "at least one seam of petroleum-bearing material." He observed that in Tarrant and Parker Counties, water wells often had "slight contamination from this oil." Such oil seeps could confound the efforts of growing communities to guarantee a reliable water supply. In 1907, for example, Decatur's attempt to drill a deep water well brought in oil with water at several levels. When Nocona drilled a city water well in 1923, it got oil production at 350 feet. Gas was equally common. USGS geologist Eugene Wesley Shaw reported, "It is commonly remarked that little puffs of gas may come from any well drilled in northern Texas, but that they have no significance"—that is, no one had found commercially attractive quantities of petroleum as a result.[7]

On occasion, locals put such gas to temporary use. The residents of Gordon, in Palo Pinto County, found enough gas in a would-be water well in the late 1880s that a venturesome soul fitted a beer keg with a pipe and burner, turned it over the well, lit the gas, and thus gave the town a night-light for some months. But, for the most part, oil or gas in a water well was cause for disappointment. As W. T. "Ole Tom" Waggoner is said to have exclaimed when a water well on his Wichita County ranch brought in oil, "I said damn the oil, I want water."[8]

Despite common disdain for oil and gas shows, from time to time well diggers encountered enough petroleum to tempt them to make a serious effort to go after oil rather than water and, in a few instances, put the oil to some use. In 1878, for example, Martin Meinsinger of Brownwood found enough oil in his water well to produce several gallons a day, which he sold as a lubricant to his neighbors. In 1890, C. M. Gearing tried drilling for oil near the Santa Fe Railroad Station in Brownwood and found several promising pay zones before losing his tools in the hole at just over 1,900 feet. Several subsequent Brownwood tests in the 1890s found gas as well as oil, but equipment mishaps downhole thwarted production. At about the same time, the Sabine Oil and Mineral Company tried a wildcat in Jack County northeast of Avis. They found oil at 117 feet but in too small a quantity to prompt development. In 1906 there was a renewed attempt to develop the field, known as the Avis or Jacksboro field, but the amount of oil produced was too small to encourage development of a market outlet. Jack County would not see commercially viable oil development until 1913, in a small field near Avis.[9]

While North Texans were both frustrated and tantalized by small amounts of oil, events elsewhere in Texas began to present oil and gas shows in a more attractive perspective. In June 1894, city leaders in Corsicana tried to drill a deep municipal water well and found oil. This time, however, the amount of oil was large enough to spur several businessmen to make something of it, ultimately bringing about development of Texas's first commercially viable oil field. Its future was assured when J. S. Cullinan, with financing from Standard Oil's Calvin Payne and Henry C. Folger Jr., built a refinery at Corsicana in 1898.[10]

Shortly thereafter, in Beaumont, a local eccentric named Pattillo Higgins persuaded several business leaders to finance a well on a hill outside town where there was a spring with a gas seep. Higgins was sure it lay atop an oil bonanza. After two abortive attempts to drill a well on the hill exhausted Higgins's cash, he succeeded in enlisting the help of Anthony Lucas, a salt-mining engineer, who, in turn, interested Pennsylvania oilmen John H. Galey and James M. Guffey in financing another test. The result proved Higgins right: a huge gusher roared in on January 10, 1901. It triggered Texas's first major oil boom, made Texas a major oil-producing state, and created an incredible amount of wealth in a hurry.[11]

The Spindletop bonanza prompted those within the petroleum industry to a profound reassessment of Texas's oil potential, and more especially, prompted them to direct capital toward exploration, particularly along the Gulf Coast. Ironically, whereas geologist William Kennedy had condemned wildcatting in this region as "idle dreams or insane notions" only a half a dozen years earlier, after Spindletop such exploration was nearly irresistible. Wildcatters brought in gushers at

Sour Lake and Saratoga in 1902, Batson Prairie in 1903, and Humble in 1905. It was tempting to think that any oil or gas seep was a sign of fabulous underground riches, especially if the seep were on or near a hill.[12]

As Texans got caught up in oil excitement, shows of oil and gas in North Texas water wells became more alluring than annoying. Thus, when Clay County farmer J. W. Lockridge found oil in a water well he had put down in 1902, he brought a fruit jar of the oil to show Henrietta banker W. B. Worsham. Worsham decided to organize a company to do more drilling on the Lockridge ranch, and he tried to interest some of the officials of Standard Oil in contributing to his venture. A fruit jar of oil was a far cry from a gusher, and the men at Standard Oil were not impressed. Worsham nonetheless got drilling started in Clay County in 1903, and by 1904 there were some seventy wells in a developing field called Petrolia. As at Corsicana, average well production was unspectacular, running at forty to seventy barrels a day. On the other hand, the field turned out to have a number of producing zones, including one as shallow as three hundred feet, and such shallow production meant it cost little to put down a well. Better yet, Petrolia production held up well over time, and the oil produced was of high enough gravity to yield profitable proportions of kerosene and gasoline.[13] In that respect, it was more attractive to refiners than heavier Gulf Coast oil.

The steadiness and high gravity of Petrolia production led some Standard Oil men, notably Calvin N. Payne and his associates, who had backed Cullinan at Corsicana, to reconsider getting into Clay County. Payne and his friends were interested in securing a dependable supply of crude for the Corsicana refinery, especially crude of a better grade than the early Gulf Coast fields produced. They organized the Clayco Oil and Pipeline Company, built storage tanks at Petrolia, constructed a pipeline from the field to the railroad, and began to drill deeper in the field. In 1907, the Navarro Refining Company, successor to their Corsicana Refining Company, brought in an impressively large gas well from a depth of about 1,500 feet. The well would change the field's future—and launch the commercial exploitation of natural gas in North Texas.

Encouraged by the well's volume and steady production, Payne and his friends joined with regional business leaders to organize the Lone Star Gas Company in 1909. Lone Star built gas lines from Petrolia to Dallas, Fort Worth, Wichita Falls, and smaller towns in North Texas. That marked the start of the North Texas natural gas industry.[14]

In retrospect, one of the remarkable aspects of the Clay County sequence of events was that someone was willing to see regional commercial possibilities for natural gas. True, natural gas had supplied northeastern urban and industrial users for over three decades by the time Lone Star Gas appeared, but for the most part, early oilmen saw a discovery of gas as nothing to celebrate. Huge quantities of natural gas were routinely flared and wasted. Drillers who brought in gas commonly let it blow freely into the air in the hope that gas flow would be followed by oil—which it often was. In practical terms, however, natural gas had to find its market in nearby areas, for the technology to send large quantities of it great distances did not yet exist. Lone Star's organizers were able to see opportunity in gas that others might have passed up. For the most part, however, as would be true for decades, oilmen wanted oil, not gas.[15]

In that context the first notable success in North Texas came as a result of Producers Oil Company (affiliated with the Texas Company) beginning exploration in 1909 on a 270,000 acre tract it leased on the W. T. Waggoner ranch, a spread spanning Wichita and Wilbarger Counties. Producers Oil began finding oil, but it did not share results of its drilling, most likely to discourage competitors from picking up leases in the area. In August 1910, it seemed to find commercial production, only to abandon the hole. Subsequently, in January 1911, a forty-five-barrel-a-day well blew in; Producers capped the well and called it a "duster"—a dry hole. Curiously, however, Producers put a wire fence around the well site and stationed a guard by it. Such secrecy may have backfired by piquing competitors' interest. At any rate, Clayco Oil Company decided to try its luck in the area, and on April 1, 1911, its Woodruff-Putnam No. 1, about two and one-half miles from the town of Electra, came in as a gusher.[16]

In an age when it took a gusher to bring an oil field to national industry attention, Clayco's well launched the first full-blown oil boom in North Texas. For drama, the Electra field's wells could not compete with the gargantuan gushers of Spindletop, but they still came in, on average, at a respectable two hundred to four hundred barrels a day. There were a number of pay zones, ranging on average from 600 to around 1,900 feet. That was shallow enough to make drilling fairly quick and wells relatively inexpensive. Like Petrolia, Electra's crude was quite high gravity and yielded a substantial amount of gasoline per barrel, so refiners were eager to buy it. Since Electra itself was only a small village in W. T. Waggoner's vast ranch pasture, those who flocked to the field either had to live there in tents or shacks or ride the train, dubbed "Coal Oil Johnny," out from Wichita Falls. And since Wichita Falls was the only town sizeable enough to serve as a management and supply center, the Electra boom turned Wichita Falls into an oil town, complete with refineries, oil field supply yards, and homegrown oilmen.[17]

By the end of 1911, over a hundred wells produced oil, and the Electra field was producing over twelve thousand barrels a day. Much of this oil was shipped out on Fort Worth and Denver City Railroad tank cars, but the first major North Texas oil pipelines were under construction by Texas Pipeline LLC and the Pierce-Fordyce Oil Association, the former organized by the Texas Company and the latter connected to Standard Oil interests.[18] Sustained, sizeable production encouraged exploration not only near Electra, but throughout North Texas. Determining where to concentrate one's search in a region thousands of square miles in extent, however, was more than a bit challenging, and the best efforts of geologists proved of only limited help.

For early twentieth-century geologists, the best source of information on underground rock formations and structure lay in what they could study and map from surface observation and learn from drillers' well logs. Trying to use either type of information in North Texas, as University of Texas geologists J. A. Udden and Drury McN. Phillips reported, was frustrating. There seemed to be an anticlinal structure trapping oil and gas at Petrolia, but if there was such a structure at Electra, it was too broad and flat to define. Udden and Phillips found Electra "quite perplexing." Worse still, subsequent geologists would find that a rock fold on the surface might not be a reliable guide to rock configurations directly underground. For that matter, there might be no surface indication of rock folds below that might trap oil. In short, what one could observe of rock structures on the surface might or might not indicate oil underground—hardly helpful to the wildcatter.[19]

As for drillers' logs, Udden and Phillips observed with chagrin that "in some respects these data are even more unsatisfactory than the data secured by examination of the surface." Most operators asked drillers to describe the rocks they encountered at specific depths, but drillers, some of whom were barely literate, had no agreed-upon terminology for what they saw, so they might refer to "red rock," "hard rock," or whatever seemed to correspond with formations they encountered while drilling other wells, sometimes in other regions. Drillers aimed to drill as rapidly as possible, only noting evidence of oil; detailed subsurface observation was unnecessary refinement. As a result, two wells in close proximity might have very different descriptions of what was underground. Beyond all these barriers to geological understanding, there was the simple fact that one did not have well logs of unexplored areas until some wildcatter drilled a well—which meant that geologists had to wait for exploration to happen.[20] Barring drilling, oil in water wells was as good a guide to finding oil as any.

Geologists found looking for oil in North Texas a tough proposition, but since the majority of wildcatters in 1911 did not hire them anyway, Electra inspired a great deal of what one might call "random drilling," done without the benefit of science. Such wildcatting began to pay off with modest finds in 1911 with the Miller field in Archer County; in 1912 three miles north of Burkburnett, in what would be known as the "Old Burk" field; and in 1913 with Iowa Park in Wichita County.[21] Unfortunately for North Texas action, on the heels of Electra development, Oklahoma upstaged Texas. In March 1912, the giant Cushing field came in, followed by another giant discovery at Healdton the following year.

Action like that would draw any self-respecting boomer in North Texas to cross the Red River and head north.

While petroleum industry attention focused on Oklahoma, a man employed in another type of hydrocarbon extraction caught oil fever—W. K. Gordon, a vice president of the Texas Pacific Coal Company. A mining engineer, Gordon developed an interest in geology, and as wildcatting brought in small discoveries in communities not far from the Texas Pacific's mines at Thurber, Gordon had a growing desire to see his company try looking for oil rather than mining coal seams. He was able to persuade Texas Pacific president Edgar Marston to send some geologists to study the Thurber area. Unfortunately, the experts saw no petroleum potential in what Gordon thought promising. Gordon, nonetheless, kept pushing for tests on company land, and in 1915 he got the company to try a test nine miles from Thurber in Palo Pinto County. The well came in for sixty barrels of oil a day—and three million cubic feet of gas, which no one wanted. No quitter, Gordon managed to get the company to back his leasing acreage in Eastland, Stephens, Palo Pinto, and Throckmorton Counties. Better yet, in 1917, when civic leaders in Ranger offered to lease him thirty thousand acres in return for Texas Pacific's drilling four tests near the town, Gordon got company agreement on the deal. In October 1917, the J. H. McCleskey No. 1 erupted in a gusher estimated at over a thousand barrels a day. Gordon's hunches paid off, and oil fever would rise to epidemic proportions in the Ranger area.[22]

The ensuing Ranger oil boom captivated popular imagination to such an extent that Hollywood would eventually use it as the setting of the movie *Boom Town*. Whether it was really as wild and wooly as portrayed, or even very different from other oil boom towns, is arguable. The boom's impact on North Texas oil, however, was undeniable. First, it gave tremendous encouragement to regional exploration. Ranger's gushers were the biggest thus far in North Texas. Better yet, wartime demand for petroleum meant a rapid escalation of crude oil prices. Even though it was necessary at Ranger to drill deeper than at Electra—beyond three thousand feet to reach the producing formation—payout

on investment from the average well was swift. While the Texas Pacific Coal Company, which would become the Texas Pacific Coal and Oil Company as a result of Ranger, had the lion's share of acreage close to the discovery, many thought those large gushers were a sign of a giant pool, extending far beyond Ranger. For that matter, oil at Ranger, and also at Breckenridge not far away, suggested that the two oil discoveries might be part of a huge producing region, perhaps 200 miles long and 125 miles wide, a vast corridor within which anywhere might be worth testing.[23]

The Ranger boom also had an impact on people that reached far beyond those who lived within sight of its gushers. The boom took place as North Texans struggled with one of the region's periodic severe droughts; the dire economic state of the surrounding countryside, after all, was what brought Ranger town leaders to lure Gordon into drilling there.[24] For farmers desperate to pay bills, an oilman's offer to lease land or buy minerals and royalties was a lifeline to survival. Farmers' sons, facing bleak futures down on the farm, could find wages in the oil field beyond anything they had ever imagined. And as exploration spread throughout the North Texas countryside, going to the oil field might be as simple as packing a pasteboard suitcase and walking down to the drilling rig in a neighbor's pasture. In effect, industrial life came to the countryside. Moreover, once World War I ended, farmers' sons headed to the oil field were joined by veterans ready for both jobs and excitement. Whether or not they spent much time or money in Ranger, for thousands of North Texas young men the action launched their transition to life in the oil fields, and having gone to the oil field, most never looked back.

A mere matter of months after production came in at Ranger, North Texas oil excitement ramped up further when a 2,200-barrel-a-day well came in near Burkburnett on the S. J. Fowler farm. For sheer drilling frenzy, Burkburnett outdid even Ranger. Wildcatters advanced on the Burkburnett townsite, leased town lots, and thickets of derricks, often as close as twenty-five feet from each other, appeared in front yards, backyards, and, in some instances, streets. As oilmen scrambled for well sites, lease bonuses skyrocketed to as much as $20,000 an acre, an unprecedented height that would not be equaled until urban Fort Worth leasing in 2008. Production from hundreds of wells swamped pipeline capacity, so operators ran oil into unlined earthen tanks; with no market for associated gas, operators simply flared it off. Such obvious waste horrified believers in petroleum conservation and would lead the Texas legislature to give the Texas Railroad Commission power to regulate oil-field practice in the interest of avoiding waste, a regulatory milestone for the industry.[25]

The Ranger and Burkburnett booms sparked successful wildcatting in dozens of locations in North Texas. Most of the finds were not very large, but since the majority of tests were the work of small independent prospectors who kept costs to a minimum, they gave many a wildcatter a start on a fortune, especially when they sold out to a large, integrated company like Humble Oil, Gulf Oil, or the Texas Company. But as prospecting paid off in other parts of North Texas, within the Fort Worth Basin, wildcatters' hopes were frustrated time after time. True, the 1920s saw the discovery and development of small fields in Cooke, Montague, and Jack Counties, but their levels of production could not compare to that of fields farther west. More often, promising shows of oil and gas lured Fort Worth Basin wildcatters to drill dry holes, making the Basin seem like an oilman's graveyard.

Activity in Wise County in the teens and 1920s offers a good example of how promising shows disappointed Fort Worth Basin wildcatters. Both Paradise and Crescent Oil, for example, tried drilling near Baker Lake in 1913-1914, a place where a gas seep had long been observed (local picnickers would occasionally ignite it for evening amusement). Both companies drilled down to around 1800-1900 feet, the depth at which there was good production in the Electra field. Both got a number of shows of gas, and Crescent may have had a show of oil, but their attempts only resulted in dry holes. Five years later, Union Oil and Gas Association tried its luck just west of Cottondale, getting shows of gas and oil at less than five hundred feet, before it gave up on the well. The Lubbock-Bridgeport Oil Company, organized by Lubbock businessmen who caught North Texas oil fever, spent almost two years drilling a well near Bridgeport to below four thousand feet; beyond a show of oil, all they reached was bankruptcy. At times, shows were encouraging enough to spur repeated efforts. In the early twenties, for example, the Keystone Oil Company drilled three tests on the Askey farm, all with promising shows, but all ultimately dry holes. Tests near Chico, Alvord, Decatur, Boyd, and Rhome had comparable results.[26]

Wildcatters in Tarrant, Parker, and Denton Counties had little better luck. In 1920, for example, Allendale Oil's test at the north end of Lake Worth got a show of oil at about six hundred feet, but that was all. Two years later, the Artz Oil Company tried drilling near Buie, getting only a small gas show at eight hundred feet. Prospectors in Parker County found no gushers but did bring in some relatively good gas wells; D. A. Upham's prospecting in 1923 brought in some gassers with production as high as eight million cubic feet a day, but gas was not what wildcatters like Upham hoped to find. As the list of disappointing Fort Worth Basin tests lengthened, no wonder oilmen shifted focus to areas more promising for oil—of which there were many in the 1920s.[27]

For wildcatters, the area around Fort Worth was disappointing, but that did not keep the city from becoming a major oil town in the 1920s. Fort Worth's position as a rail hub helped in this regard. Long a cattle shipment and packing center, Fort Worth was served by six major railroads. With respect to oil, the most important railroads were the Fort Worth and Denver City and the Texas and Pacific. The Fort Worth and Denver allowed oil from the Wichita Falls area to reach Fort Worth refineries, the earliest of which were built by Gulf, Pierce-Fordyce, and Magnolia; the Fort Worth and Denver also took oil-field equipment and lumber out to the fields. Fort Worth did not compete with Wichita Falls as an oil center during the first stage of North Texas development, but once Ranger came in, advantage shifted from Wichita Falls to Fort Worth. The T and P served the towns in the Ranger area and westward into the Permian Basin, where huge oil fields would be discovered in the 1920s. As a result, the T and P assured Fort Worth's preeminence as an oil-field supply center for a vast area after 1918; by 1920 there were fifty-two oil-field supply companies with Fort Worth offices. And, as pipeline construction allowed an increasing volume of crude oil to reach the town, there was ample incentive to build local refineries. By 1924, some nine refineries, most small scale, served a growing urban market.[28]

Because Fort Worth was "Cowtown" before it was an oil town, the city also had the kind of financial and business infrastructure that could support a growing oil community. West Texas farmers and ranchers had long shipped cattle to Fort Worth, used Fort Worth banks for loans and business operations, and bought farm and ranch supplies in town. As a regional business center, Fort Worth also had office buildings and a white-collar workforce, making it attractive as a management center. Escalating regional oil activity encouraged construction of additional multistoried buildings in the 1920s—the W. T. Waggoner Building, the Farmers and Mechanics National Bank Building, and, inevitably, the Petroleum Building, designed by Wyatt C. Hedrick in a style very similar to "petroleum buildings" in Corsicana and Midland. Proximity to North Texas oil action, railroads, and financial and business infrastructure made Fort Worth an attractive regional management center for large oil companies, and by the mid-1920s, many had Fort Worth offices, most notable among them Gulf, Humble, Sinclair, Continental, and Pure.[29]

The features that made Fort Worth attractive as a base of regional operations for major oil companies also encouraged the development of a large independent oil business community. In fact, the presence of major company offices gave independents the opportunity to interest majors in their projects—whether raising "dry hole money" that majors might offer to wildcat, obtaining farm outs of major company acreage, or selling and trading productive leases. Among the most prominent independents of the 1920s were J. D. Collett, E. A. Landreth, William H. Dunning Jr., Charles F. Roeser, Tolbert T. Pendleton, C. O. "Ted" Collins, Luther C. Turman Sr., Joseph C. Maxwell, George B. McCamey, and Burton F. Weekley. These independents moved beyond regional exploration to try their luck in the Permian Basin and elsewhere. In

Fort Worth, they congregated in the lobby of the Westbrook Hotel, where a statue of a scantily clad maiden holding a torch presided over the crowd. It became a tradition to pat the rump of this "Golden Goddess" for luck when launching the next deal. To judge from the amount of petroleum found by Fort Worth independents, the Golden Goddess was mightily accommodating.[30]

During the early 1920s, Fort Worth also hosted what amounted to an independent oil underworld, a horde of mail-order oil promoters looking to reap cash from investors hoping to get rich quick. Some promoters did put investor dollars to work, picking up acreage and drilling tests. Too often, however, all investors got was fancy promotional literature, promising riches beyond their wildest dreams. Promoters often bid for investor attention by trying for celebrity. Dr. Frederick A. Cook, for example, claimed renown as an Arctic explorer, maintaining he had reached the North Pole before Admiral Peary. S. E. J. Cox, nicknamed "Alphabet Cox," created the persona of a dashing young aviator leading life in the fast lane. Charles Sherwin and Harry H. Schwarz manufactured a celebrity by finding an elderly drifter with a white beard named Robert A. Lee and presenting him as a nephew of the Civil War general. The "General," they told investors, wanted to lead his investors to wealth by drilling wells in "the heart of Texas"—which turned out to be Denton and Tarrant Counties.[31]

These promoters and many others like them turned Fort Worth into the epicenter of mail-order oil promotion in the early 1920s. They kept print shops busy turning out tons of promotional literature that was sent throughout the United States. Hundreds of bags of mail returned, stuffed with envelopes enclosing investor dollars. The golden age of mail-order oil promotion wound down in 1922–1923, however, when postal authorities began to haul promoters into court. Some of the higher profile promoters, including Cook, Cox, and the General Lee circle, ended up doing hard time, thus thinning, but by no means eliminating, the city's mail-order promotion clan; those remaining found ways to adapt their strategies to stay just within the boundaries of the law. But after the East Texas Field came in, and oil prices dropped to ten cents a barrel, even the craftiest promoters had a hard time persuading investors that oil would make them rich.[32]

Fort Worth's oil fever of the twenties would give way to the less hectic routine of an established oil community in the 1930s. Weathering the Depression, its oilmen would continue to open up and extend hundreds of oil fields, returning their focus to North Texas, the Permian Basin, and southeastern New Mexico. Thus, Fort Worth stayed an oil town—ironically without a single derrick or pumping unit within city limits. Its wildcatters would continue to pat the rump of the Golden Goddess, but they went looking for oil somewhere else.

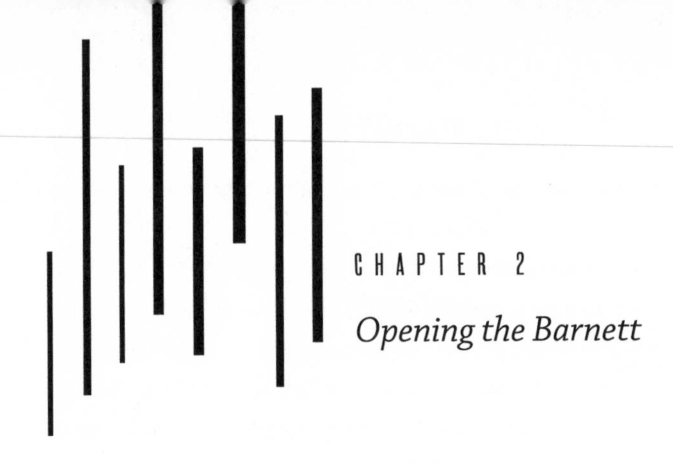

CHAPTER 2
Opening the Barnett

Even if one thought that North Texas was not past its peak as an oil-producing region in 1945, one would not have anticipated much profitable exploration and development in the Fort Worth Basin. Nor would one have predicted a fortune could be made producing the region's natural gas. From a conventional industry perspective, North Texas was a mature producing region in which deeper drilling might mean some modestly attractive finds, but exploration dollars might better be spent elsewhere. Thus, regional development would have its ups and downs in the half century after World War II, when the industry as a whole experienced periods of boom and bust, but better-funded wildcatters, like the major companies, directed their attention elsewhere. To open up production from the Barnett Shale, a formation whose gas had not been profitable to exploit, it took a wildcatter with an unconventional exploration strategy—Houston independent George P. Mitchell.

The war's end saw the American economy enter an extended period of tremendous growth, a nationwide boom that generated skyrocketing demand for oil and natural gas. In 1946, when wartime price controls on crude oil ended, oil prices soared. Oilmen had postponed many projects during the war because of shortages of materials and labor; now they hurried to take projects off the shelf and put thousands of veterans to work in the field. Rising crude oil prices let prospectors in North Texas, as elsewhere, take on the expense of deeper drilling to wildcat in a new range of formations, but they might also let oilmen make money on the sorts of ventures typical of the mature producing region, drilling relatively inexpensive, shallow wells to tap smaller pools. Higher prices made up for modest yields. Accordingly, both major companies and independents took another look at petroleum possibilities in the Fort Worth Basin. True, its tantalizing oil shows in the past had led to drilling many a dry hole, but deeper drilling might bring in profitable production. Moreover, applying new technology, in the form of seismic work in exploration and well stimulation through acidizing and hydraulic fracturing (fracking), might mean better chances for paying wells. And in an area where tests for oil had often only brought in gas, there was yet another encouraging new development: a postwar surge in demand for natural gas.[1]

Though used as fuel for decades, before 1945 an enormous amount of natural gas went without a market, largely for lack of pipelines capable of shipping large quantities over long distances. In the late 1920s, improved pipeline-welding technology began to overcome this barrier. In the 1930s, long-distance gas pipelines connected Texas Panhandle fields to various midwestern cities, but most natural gas was still consumed close to producing fields or, when produced with oil, simply flared at the wellhead. Wartime demand for petroleum, however, encouraged shipment of southwestern natural gas to fuel factories and steel mills in places like Ohio and Pennsylvania. Because natural gas was so inexpensive, demand for it accelerated after the war, and gas companies proliferated as Texas oilmen stepped up to the plate to meet that demand. The gas producer thus found it much easier to sell gas but still would not receive high prices for it. In particular, gas going to the interstate market fetched prices held at very low levels by the Federal Power Commission, and until the 1960s, federal price controls also helped hold back intrastate gas prices. For the prospector, the key to making money from gas production lay in finding something big.[2]

Under these circumstances, when oilmen looked for petroleum in the Fort Worth Basin, they continued to be more interested in oil than gas. This was especially true for major companies like Cities Service and Continental Oil, companies with crude-hungry refineries. As the majors bankrolled deeper drilling, they began to bring in commercially attractive amounts of oil in places where earlier finds had been meager. Fort Worth Basin action was initially hottest in the north, in Clay and Montague Counties, but by 1947, it had moved south into Wise, Jack, and Parker Counties, where drilling into the Bend Conglomerate formation, which was about a mile deep, brought in successful oil wells. When exploration picked up in these counties, however, oilmen began finding more gas than oil. In 1950, Continental Oil's Wise County test, Bertha Flowers No. 1, seven miles south of Bridgeport, opened a giant gas field producing from the Bend Conglomerate. Seven years later, the Railroad Commission of Texas would group the field opened up by the Flowers well with twenty-eight other gas fields to form the Boonsville (Bend Conglomerate) gas field, stretching over 390 square miles in Wise, Jack, and Parker Counties.[3] In terms of industry significance, a find like this could not compete with the enormous gas fields that would open up in places like the Permian Basin in the 1960s, but it was clear that the Fort Worth Basin was no longer an oilman's graveyard. How profitable it might be remained to be seen. In that context, a young George P. Mitchell would play a leading role.

The story of George Mitchell's early life may remind one of classic American fiction. His father Savvas Paraskevopoulos left Greece, where he had been a shepherd, to come to the United States in 1901. Illiterate and unskilled, Savvas found work in railroad construction. Unable to pronounce his name, an Irish paymaster declared, "You're Mike Mitchell from now on, and that's it!" Several years later, Savvas and his cousin set up a pressing shop and shoeshine stand in Galveston, Texas. His son George was born in Galveston in 1919.[4]

The Mitchell family lived above the pressing shop until 1933, when the death of George's mother and accidental injury of his father split up the family. George went to live with an aunt while he finished high school; his older brother Johnny graduated from Texas A&M University and set up an oil well workover firm. Having spent a summer helping Johnny, George was encouraged to pursue a career in petroleum engineering. He put himself through Texas A&M by waiting tables, delivering laundry, and selling candy bars while taking as many courses as he could in both petroleum engineering and geology. When he graduated in 1940, he took a job with Standard Oil. At that time, major companies commonly assigned novice geoscientists hands-on hard work in the field, which was George's lot in his first months on the job. Enduring nights working in rain and sleet, he had some second thoughts about his vocation: "I thought, my God, if I thought I'd have to do this all my life, I'd go jump off this damned rig right now."[5]

Uncle Sam saved George. Once the United States entered the war in 1941, George, now an Army Reserve officer, was assigned to the Army Corps of Engineers and spent the war managing large construction projects. At the war's end, George and Johnny partnered with Houston oil broker H. Merlyn Christie to found Oil Drilling, Incorporated. They rented a tiny office in the downtown Houston Esperson Building, acquired an antiquated drilling rig, and hoped for business to come their way. While Johnny and Christie worked at putting together deals, George used his training in geology and engineering not only to work up projects for the firm but also to evaluate those being considered by local investors. Many of the projects focused on finding natural gas rather than oil, a focus George would keep throughout his career. His consulting helped the firm build a circle of investors, including oilmen like R. E. "Bob" Smith, Louis and Henry Pulaski, and Morris Rauch, as well as a number of other prominent businessmen. It also led the firm to some significant gas discoveries in Gulf Coast fields. But the firm's big breakthrough came in 1952, when Louis Pulaski asked George to look at a project proposal sent to him from a most unlikely source: Pulaski's Chicago bookie.[6]

The project originated with geologist John A. Jackson and North Texas driller Ellison Miles, who wanted to drill a well on three thousand acres of the Hughes Ranch in Wise County. Perhaps because of Wise County's checkered industry track record, they had not found investors. By the time the proposal reached George, it had been in circulation for two years, and, as he put it, "was pretty tattered around the ears." Upon examination, however, George concluded that Jackson had identified a major stratigraphic petroleum trap, and that the project was worth trying. Going forward, the Mitchells, Christie, and their investors drilled their D. J. Hughes No. 1. It came in at over a million cubic feet of gas per day, prompting the Mitchells to scramble to lease as much acreage as possible. Within ninety days, they managed to lease three hundred thousand acres in Wise, Jack, and Parker Counties, at an average price of three dollars per acre. The leases would be the cornerstone of the Mitchells' future North Texas production.[7]

Following the Hughes Ranch discovery, the Mitchell group drilled ten more wells without a single dry hole, a stunning record of success by prevailing industry standards. Like the Bertha Flowers well, the Mitchell wells produced gas from the Bend Conglomerate at depths ranging from five thousand to six thousand feet. Part of this drilling success was the result of George's decision to apply the relatively new technology of hydraulic fracturing, "hydrofracturing," as it was called, to enhance well production. This technology, better known as "fracking," involves injecting fluid bearing a proppant material, like grains of sand, at extremely high pressure into the rock around the wellbore. Under high pressure, the fluid pressure hydraulically creates a fracture; the myriad of fractures in the rock stay open because the proppant keeps them from closing once pressure subsides. Oil and gas molecules can readily move through the sand-filled openings that fracking creates. Oddly enough, although fracking had been in use for four years, major companies had not used it in Mitchell's area. In fact, when he tried going back into wells lying within the partners' acreage that the majors had abandoned as dry holes, fracking often turned these supposedly dry holes into commercially viable gas producers. In any event, Jackson's geology paid off. The Mitchell group's acreage lay over an enormous reservoir of natural gas.[8]

For the Mitchell group to make any money from their fabulous find, much had to be done. Of course, they had to find a buyer for their gas, but wells had to stay shut in until the infrastructure to handle gas was in place. Gathering lines were necessary to take gas from the wells. Rich in liquids like propane and butane, the gas would then need processing to extract the fluids and make the gas ready for the pipeline. The likeliest regional buyer was Lone Star Gas, but that company proved unwilling to give the group any concessions in a purchase, offering a rock-bottom price of eleven cents per thousand cubic feet (Mcf). Moreover, Lone Star insisted that it would extract, and thus profit from, the liquids in the gas. George's reaction was to tell Lone Star, "To hell with you. We're going to build our own [processing] plant."[9] Fortunately for George, an alternative buyer appeared, Natural Gas Pipeline of America (NGPL). With a trunk line to Chicago, NGPL was shopping for natural gas to serve its urban market. The large reserves Mitchell's group found were precisely what NGPL needed. NGPL not only offered the group a premium price for the time—thirteen cents per Mcf—and a twenty-year contract, but advanced them $7 million to put in infrastructure and drill more wells. In December 1957, Mitchell's firm began delivering gas to NGPL, the beginning of a forty-year relationship in which Mitchell's wells would come to supply 10 percent of Chicago's gas.[10]

By the time NGPL began receiving the Mitchell group's gas, the Wise County venture had absorbed a huge amount of capital. Many early investors had grown reluctant to sink ever more cash into a deal that, until the end of 1957, produced no return. Christie and the Mitchells had to borrow money in order to do the drilling necessary to hold leases. Fortunately, Harold Vance, one of George's former Texas A&M engineering professors, was vice president for energy at the Bank of the Southwest and helped them get a loan. The Mitchell group also needed to go ahead with building a gas processing plant, and here they needed not only cash but also expertise. They made a deal with Warren Petroleum, a leading gas processor, by which Warren would direct construction and operations in return for a 20 percent interest in the future plant. Warren was also able to negotiate a loan from First National Bank of Chicago to get construction underway. The processing plant, built at Bridgeport, would be a key element in George's ability to make money during the downtimes of the 1960s and 1970s, when federal regulation held interstate natural gas prices at unreasonably low levels. Low prices kept the plant's raw material, whether from George's wells or those of other producers, ridiculously cheap, while escalating demand for natural gas liquids meant rising prices for what the plant produced. Thus, profits from processing would do a great deal to offset flat income from the sale of natural gas.[11]

It is ironic that just as the Mitchell group began to sell their North Texas natural gas, the independent sector of the US petroleum industry entered a fifteen-year period of hard times, during which many oilmen found it challenging to stay in business, let alone make substantial profits. Those producing oil were, in effect, competing in a global market with Middle Eastern producers able to produce crude oil at a fraction of the cost of American production. Crude oil prices stayed below three dollars per barrel, and to try to keep them from sliding lower, the Railroad Commission put stringent limits on how much Texas wells could produce. But for gas producers, the economic challenge came from Washington rather than overseas, in the form of interstate price ceilings set by the Federal Power Commission (FPC). Though prices for both oil and gas stayed low, costs of materials, labor, equipment, and services rose, creating what the industry called a "cost-price squeeze." In order to be successful, an independent had to craft a strategy around bleak economic reality.[12]

To offset industry doldrums, George expanded profitable processing operations, diversified both geographically and economically, and over time consolidated operations into his own hands, progressively buying out investors and partners. In order to increase the amount of natural gas supplying the Bridgeport plant from wells of other gas producers, George bought Walter Davis's Southwestern Pipeline network in 1963. By the early 1970s, development of smaller mobile processing units allowed him to take processing into small North Texas gas fields. He also built reserves outside North Texas, drilling inside and offshore Galveston city limits, and he made important oil and gas discoveries in Limestone, Polk, and Aransas Counties. Moving beyond petroleum, he went into real estate by launching a planned community at the Woodlands, north of Houston. And over time, his firm evolved—from Oil Drilling, Incorporated, to Christie, Mitchell and Mitchell in 1953; to buying out Christie and forming Mitchell and Mitchell Gas and Oil Corporation in 1962; to eventually buying out his brother Johnny and all other investors. Finally, in 1972, after urging from bankers worried about the amount of his debt, George formed the public company, Mitchell Energy and Development Corporation. Going public would facilitate raising much more capital when industry fortunes turned around, which they began to do in 1973.[13]

The seventies would go down in national history as a time of energy crisis, of natural gas shortages that closed schools and factories, and of gasoline shortages aggravated by the Arab oil embargo that resulted in long lines at filling stations.[14] For American independent oilmen, however, the seventies meant rising petroleum prices and revived opportunity, culminating in an all-out industry boom at the decade's end. As in other regions, after years of doldrums, leasing and drilling picked up in North Texas and accelerated most sharply in the Fort Worth Basin. By September 1974, the *Oil & Gas Journal* reported that, in terms of drilling, the Basin "has to be one of the busiest places on the map."[15] Responding to dramatic rises in gas prices on the intrastate market, many prospectors shifted exploratory focus from oil to gas. At the end of 1974, for example, when interstate gas still went for less than eighteen cents per Mcf, gas produced and sold in Texas could go for as much as $1.46 per Mcf. For prospectors still focused on oil, there were many areas in North Texas where exploration brought in small discoveries that could nonetheless return solid profits—especially for independents who could keep operating costs low. As one independent would later remark, "This is rabbit country as opposed to elephant country. We're looking for small structures, not giant fields. But it's

that type of production that's bread 'n' butter for most independents."[16] Indeed, by the 1970s, North Texas was independent turf, its targets far too small to appeal to major company elephant hunters. But even as exploration picked up, exploratory objectives did not include the Barnett Shale. Though prospectors had ample evidence that hydrocarbons were present in the Barnett, prevailing opinion held that it was "nonprospective as a hydrocarbon reservoir."[17]

The more venturesome of Fort Worth independents, not content to drill close to home, embarked on a wide range of projects. New independent firms proliferated, ready to take on ambitious exploration agendas. American Quasar, for example, founded by Ted Collins, David McMann, and Wilbert Fultz, and former TCU football star Dick Lowe began by picking up North Texas properties sold by Tenneco but went on to make major discoveries in the Rocky Mountain Overthrust. The number of drilling and oil field service contractors surged upward, while older firms like Eddie Chiles's Western Company did record levels of business. Chiles gained local immortality with a 1973 advertisement suggesting, "If you don't have an oil well, get one—you'll love doing business with Western." Later in the seventies, Chiles began radio broadcasts attacking Washington's energy measures, announcing to listeners, "I'm Eddie Chiles, and I'm mad!" Soon bumper stickers throughout the southwestern oil patch proclaimed, "I'm mad, too, Eddie." After years of idling, the petroleum industry was once again running in high gear. And many an economist assured energy observers inside and outside the industry that oil and gas prices had nowhere to go but up.[18]

As oil and gas prices rose, Mitchell Energy lost no time in adapting strategy to industry revival. It ramped up its drilling program, looking for oil as well as gas, and to ensure it could get rigs when it needed them, the company set up a drilling subsidiary, MND Drilling. Some wells went down in unusual locations. In 1981, for example, regional drought lowered water levels enough in Lake Bridgeport to lead the company to drill on leases from the Tarrant County Water Board, which it had held in inventory since 1954, a project resulting in six wells. It went back into oil-producing leases near the Wise County town of Alvord, leases under secondary recovery by waterflooding, and experimented with using liquefied petroleum gases for tertiary recovery. It launched a tertiary recovery project using carbon dioxide, a by-product of the Bridgeport gas processing plant. Going outside the region, Mitchell Energy tried drilling in the Austin Chalk, where finding natural rock fracturing was key to successful production and where some prospectors were experimenting with the new and expensive technology of horizontal drilling. And as more and more gas came on line throughout Texas, George expanded processing operations—always a solid moneymaker for the firm.[19]

George's business strategy, however, was influenced as much by changes in federal energy policy as by changes in industry fortunes. In 1978, Congress passed the Natural Gas Policy Act, the most important measure in President Jimmy Carter's energy program. Among its many provisions, the act extended federal control of natural gas prices to intrastate markets, a blow to many Texas producers who had enjoyed the higher prices of intrastate sales. But the act also mandated higher prices for gas found and produced below fifteen thousand feet or from problematic "tight sand" formations—formations where it was more difficult than elsewhere for gas molecules to move through rock to the wellbore. In such formations, only applying sophisticated fracking technology was likely to yield commercially attractive amounts of natural gas. Going after tight sand gas was riskier and more expensive than finding gas in less challenging formations, hence the federally mandated higher prices, which were supplemented by a tax credit. The new federal policy played into a staple of George's operating strategy: using fracking to bring in commercially successful wells. So if Washington was willing to reward those who went after tight sand gas, George was ready to take advantage of federal incentives.[20]

Some of the tight sand gas contained in the Barnett Shale lay right below Mitchell's North Texas leases, but going after it would not be easy. For years, oilmen had gotten some gas when they drilled into or through the Barnett Shale, but the Barnett never gave up enough gas to make commercially viable wells. Covering much of the Fort Worth Basin, the formation lies below seven thousand feet, with upper and lower levels separated by a layer of Forestburg Limestone. Technically not a shale (because it contains little clay) but a mud rock, the Barnett contains a high percentage of organic matter, having the potential to generate huge quantities of hydrocarbons. The mineral grains within this rock, however, are so densely packed together that the Barnett has exceptionally low porosity and permeability. In contrast to a conventional gas-producing formation, where gas molecules may move readily through rock into a wellbore, it might take millions of years for petroleum molecules to move from the Barnett into a more permeable rock formation that would allow profitable levels of production. Simply drilling into the Barnett does not release the gas around the wellbore. As one scientist put it, the Barnett is "as hard as tombstone granite."[21] One could believe it had about as much production potential.

Notwithstanding the tightness of rock like the Barnett Shale, natural geological upheavals can break or shatter it, creating fissures or fractures through which hydrocarbons can migrate into more porous and permeable rock formations from which they can be readily extracted. In fact, James Henry, one of George's geologists, suggested that the Barnett Shale might be the source rock for some of the longest-lasting petroleum production in the Fort Worth Basin. If Henry was right, Mitchell's leases might be over a veritable petroleum bonanza, something George and his North Texas exploration manager, Bob Gardner, were quick to recognize.[22] There had been so little study of the Barnett, however, that drilling on the basis of Henry's suggestion alone would be hard to defend.

But geology was not the only element bearing on a decision to test the Barnett. George already had leases and infrastructure where he might try a Barnett wildcat. If he brought in a commercial level of Barnett gas production, he could sell it directly to NGPL. As tight sand gas, it could command a higher price than gas sold under the original NGPL contract. In any event, by 1981, NGPL expressed concern that wells in production from the Bend Conglomerate for a quarter century might falter as sources of gas supply. George needed to address that concern by finding more gas. In southeastern Wise County, he already had wells producing from the Caddo Limestone; therefore, if he drilled a disappointing Barnett test, he could always plug back in to production from that source. During the mid-1960s, Phillips Petroleum Company tried some tests near those Caddo wells, losing interest when it got nothing big. Perhaps Phillips gave up too soon. George did not want to overlook what might be a good play in the future. As he put it, "First off, it would be stupid, and secondly it'd be embarrassing, and I don't want to be either."[23]

Early in 1981 a projected development well on an expiring lease near Rhome in Wise County offered the chance to go beyond the Caddo Limestone to the Barnett and the Viola Limestone underneath it. At George's insistence, the well, C. W. Slay No. 1, was completed in the Barnett in late September. After a small frack, it tested for only 246 Mcf of gas per day, a rate that was not only disappointing but so low that NGPL did not have to take its gas; thus, Mitchell Energy had to bear the cost of connecting Slay No. 1 to a gathering system. Nonetheless, Mitchell Energy filed for a new field discovery. The C. W. Slay No. 1 was thus the discovery well for the Newark East (Barnett Shale) Gas Field.[24]

Though C. W. Slay No. 1's production was not impressive, George was not about to give up on the Barnett, as his geologists and engineers soon learned. The initial problem they faced was lack of information about the formation. Was it like the Marcellus Shale in Appalachia? Were there identifiable structures within it? Would drilling in areas with many natural fractures mean better wells, as was true in the Austin Chalk? How would it respond to different types of fracking? No one, including George's own scientists, had bothered to collect much information about a formation generally dismissed as commercially unproductive. Mitchell geologists, led by John Hibbeler, Dan Steward, and Christopher Veeder, decided to gather more information by deepening existing wells and buying seismic data; they began to put together maps and studied core samples. Meanwhile, George's engineers experimented with different types of fracking substances, including carbon

dioxide foam and nitrogen-assisted gels that used massive amounts of fluid and sand. Discouragingly, with study and experiment, it became clear that the Barnett was not like tight rock in other places. For example, in the Austin Chalk, drilling near naturally fractured areas was one of the secrets of success. Where Mitchell drilled in the Barnett, natural fractures were healed, closed with calcium carbonate, and did not allow hydrocarbon migration. Similarly, the massive fracks that worked well in other fields had unimpressive returns.[25] What worked elsewhere failed in the Barnett.

By the mid-1980s, it was clear that although Barnett wells had to be fracked to yield acceptable production, George's engineers needed to find a better fracking protocol than they were using. To do so, they needed to experiment with more wells drilled to the Barnett. Unfortunately, industry economics turned against them. Since the late 1970s, higher gas prices encouraged prospectors to go out and find a super abundance of natural gas, which swamped the domestic market. Worse yet, as long-term contracts locked in higher gas prices, demand fell when industrial users switched to cheaper coal. The result was a "gas bubble," a surplus that led natural gas purchasers to try to renegotiate contracts and back away from new purchases. At the same time, record-high oil prices at the beginning of the decade led to a global oil glut, pushing prices down until they crashed in March 1986, dropping to under ten dollars per barrel. Caught in the worst industry downturn since the Great Depression, many oil companies trimmed exploration and development back to the bone for survival. Mitchell Energy was lucky enough to have a long-term contract with NGPL to sell gas, thus averting cash flow crisis, but the firm cut back expensive projects like drilling to the Barnett. Thus, in 1986, Barnett completions were only 1 percent of its drilling program.[26]

Undaunted by bleak economics, George did not give up on the Barnett Shale. Instead, in 1985, he expanded leasing to pick up one hundred thousand acres that might one day be productive in Denton County, and he continued leasing there and in Tarrant County until 1990. Then, working within a core area of southeastern Wise, southwestern Denton, and northwestern Tarrant Counties, George's staff set about deciphering the geology of the Barnett and bringing down the cost of wells to tap it. They were able to get some funding help from the Gas Research Institute for experimentation, but most of the cost was borne by the firm, which generated controversy; some of George's own employees thought he was wasting money. Even optimists could not be sure of the long-run return of pursuing the Barnett; they could only guess at how much gas the rock held.[27] Continued work, however, did succeed in cutting costs of drilling. Better understanding of the formation let geologists avoid formational quirks that meant expensive drilling problems, and more drilling let engineers reduce the time it took to drill a Barnett well by roughly half, to eleven days. But by the mid-1990s, there still had been no dramatic breakthrough in well productivity or overall well cost. As Mitchell staff member Dan Steward observed, it took a long time "to figure out what we were doing wrong."[28]

Changing completion technology proved key to raising productivity and lowering costs. In the mid-1990s, Mitchell staffers George Jackson, Tim Addison, and Nick Steinsburger began a thorough reexamination of completion practice, which at that point often amounted to as much as half the total cost of drilling a Barnett well. They had been using some acidizing before fracking, nitrogen after fracking, and a massive amount of heavy gel fluid carrying a large amount of sand to fracture the rock. Upon reexamination, they discovered they could reduce costs by eliminating both acidizing and use of nitrogen. The biggest breakthrough in both well productivity and cost reduction, however, came when they decided to try what was called slickwater or light sand fracking—using a frack fluid mostly composed of water with added surfactants and a greatly reduced amount of sand. This new frack approach, used by Union Pacific Resources on the Cotton Valley tight limestone, could result in a much wider area of shattered rock around the wellbore, and therefore drainage of a much greater amount of petroleum into the wellbore.[29]

Using a slickwater frack was a daring departure because it went dead

against conventional industry thinking, which held that water should not be used to stimulate production from shale. In shales originally containing a great deal of clay, applying water in a frack would make it harder for petroleum molecules to move to the wellbore. George's engineers and geologists, however, had worked long enough to take the risk that conventional thinking was wrong for their core area. Moreover, if water would work, there would be a savings of $150,000 or more in completing a well. In 1997, they began using slickwater fracks, and their gamble paid off: they gained two productive wells and saved $200,000 per well.[30] Because slickwater fracking worked well, two other developments took place that greatly increased Barnett gas production—one that could have been anticipated and one that was unexpected.

The development that followed logically from greatly reduced frack cost was the ability to begin tapping the upper part of the Barnett Shale, above the Forestburg Limestone, for production. With a view to limiting costs, George's engineers had been fracking only the lower Barnett when they completed wells because, since it was thicker than the upper level, they reasoned it was more likely to yield more gas. If one had to choose, the lower Barnett seemed the better target, but the savings from using a slickwater frack made it affordable to frack both upper and lower formations, thus unlocking additional production from what had been purposely passed over. Even more remarkable, a slickwater frack of the lower Barnett could result in good production from the upper Barnett even without additional stimulation of the latter. In short, both parts of the Barnett were now profit-makers. One notable well, the Blakley Estate No. C-1, came in at 1.3 million cubic feet of gas per day; in a year, the well yielded 235 million cubic feet of gas. No wonder that, by the end of 1998, Mitchell engineers were exclusively using slickwater fracking.[31]

Reliance on slickwater fracking was doubly encouraged by a completely unforeseen development: slickwater fracking's impact extended beyond the well being stimulated to older adjoining wells. Much to Mitchell engineers' surprise, wells with settled production near the well being fracked suddenly started producing as much as eight hundred Mcf more gas per day—at no additional cost to the company! Not surprisingly, in 1999, Mitchell Energy decided to drill many more development wells and rework older ones with slickwater fracks. Often production from reworked wells was far greater than it had been initially. As Dan Steward summed up, "Our rate [of production] just took off like a rocket."[32] After seventeen years and an investment of at least $250 million, George's faith in the Barnett Shale was vindicated.

Once the results of slickwater fracking were in, it was clear Mitchell Energy had opened something big. How big was a matter of speculation. In order to construct its Barnett development strategy, in 1991 the Mitchell geologists worked on estimating how much gas was contained in the formation within its core area and arrived at an estimate of roughly fifty billion cubic feet of gas per square mile. In 1998, former Chevron geologist Kent Bowker joined George's staff in testing for gas in the Barnett in Johnson County. After some months of study, Bowker concluded that the company's estimate was far too conservative—that there might be two or three times more gas locked in the rock. For all the remarkable increase in production from slickwater fracking, George might, in fact, be recovering only 6 percent of the gas in place. Bowker also argued that the Barnett would be productive far beyond George's core area into large parts of Tarrant, Parker, and Johnson Counties. With a vision of a far larger bonanza than it could have anticipated, Mitchell Energy began aggressive leasing in those counties, with only one limitation: George Mitchell himself forbade leasing within the Fort Worth city limits. He had experienced the aggravation of town lot leasing in Galveston three decades earlier and had no wish to repeat it on a grander scale.[33]

At the end of 2000, Mitchell Energy's exploration and development of Barnett leases were in top gear. The company had nine rigs working constantly, drilling on average two wells a month. It was reworking between seventy-five and ninety wells annually, thereby adding an additional five hundred thousand cubic feet of gas per day to its production. To ensure ample water for fracking, it put down water wells capable of yielding five thousand barrels of water per day. Its gas production was so great that it had had to extend its marketing to East Texas and the Houston area. And by that time, not only did it have independent competitors tapping the Barnett, but George's spectacular success had captured keen industry attention. In particular, Oklahoma City-based Devon Energy, with a similar focus on gas, began talking with the company about a merger in the spring of 2001, a merger that was announced on August 14 of the same year. Devon purchased Mitchell Energy's assets for $3.1 billion and assumed $400 million of Mitchell's debt. By the time the merger was official at the end of January 2002, daily Barnett production from Devon's newly acquired leases was 365 million cubic feet—and what would become an all-out boom was underway.[34]

All this was largely the result of George Mitchell's tenacity and willingness to take the sort of risks that flew in the face of conventional thinking. How could so much gas be produced from a rock as unpromising as Barnett Shale? How could using a fracking fluid composed mostly of water be superior to ordinarily used gel? For that matter, how could a deal worth considering come from the hands of a Chicago bookie? Most importantly, as Dan Steward argues, George Mitchell was a visionary who believed that the future of energy lay not in oil but in natural gas.[35] Vision, however, is unproductive without the courage to act on it and the determination to keep going against the odds. In the mid-1990s, most Fort Worth independents either condemned Mitchell for sitting on a huge lease inventory without substantial production or saw him as deluded. As Ken Morgan summed up the oil community verdict, Mitchell was a "good guy, worked hard, tried to make it, made plenty of money . . . but this [going after the Barnett] is folly."[36] Larry Brogdon likened Mitchell's determination to unlock the Barnett to the tenacity of a terrier:

> I don't know if you've ever seen a Boston terrier. . . . You give 'em a towel, and they'll grab it in their mouth, and you play tug of war with 'em . . . and they won't let go. That's the way he was about the Barnett.[37]

Mitchell's stubborn faith in the Barnett paid off beyond what anyone could have predicted. As Charles Moncrief observed, "He just revolutionized the oil business. That's all there is to it. Just simply that man."[38] Brogdon summed up Mitchell's achievement more graphically:

> We can never repay George Mitchell for what he did. . . . He stayed with it when nobody else would. . . . The city of Fort Worth ought to have a huge statue of [him], like what they've got when you drive down on [Interstate] 45 going to Houston, you see a big statue of Sam Houston. There ought to be one like that for George Mitchell for what he's done, not only for this area but for what it has done for our country . . . and what it's going to do for the world.[39]

Like John D. Rockefeller, George Mitchell had an enormous impact on the global petroleum industry. But as it was with Rockefeller, what he did in showing how one could produce petroleum from previously untapped and untappable source rock would generate unforeseen controversy as well as enormous wealth.

CHAPTER 3

Elephant in Rabbit Country

DURING THE SEVENTEEN LONG YEARS THAT MITCHELL Energy spent experimenting with the Barnett Shale, the company's efforts did not go unnoticed by others in the industry. From time to time, oil and gas companies, both great and small, tried their luck in the Barnett but lacked either the inclination or the resources to persevere in tapping the formation. Even if one did not believe that George Mitchell was deluded, there were good economic reasons to pass up testing his faith in the Barnett. For large companies, bigger targets elsewhere were far more easily justified in exploration budgets. For smaller players, though Fort Worth Basin leases were not expensive, massive gel fracks used to bring in production were pricey, especially in the light of the low natural gas prices of the late 1980s and 1990s: small independents could not afford to imitate Mitchell Energy. Mitchell's discovering how to use slickwater fracking on the shale, however, changed everything, especially for Fort Worth independents, who began to scent a bonanza in their backyard. A handful of small independents played a critical role in expanding Barnett wildcatting far beyond Mitchell's core area, and thereby launched a shale boom.

The history of major and large independent companies' explorations of the Barnett before 2000 demonstrates how easily the elephant hunters—those out to make giant discoveries—gave up on what seemed to be rabbit country. Perhaps attracted to the Fort Worth Basin by ample Boonsville Bend (Bend Conglomerate) gas production, Phillips tried its luck in southwestern Denton County in the mid-1960s. It completed one gas well (which it shut in), drilled a test through the Barnett to the Ellenburger formation—only to plug and abandon it—and then gave up on the area. A quarter century later, in 1991, Oryx Energy, the exploration unit spun off from Sun Company, Inc. (later Sun Oil), leased in Montague and Tarrant Counties and tried a Barnett Shale wildcat in Montague County, hoping to get oil. Disappointed by meager production there and a dry hole in northwest Tarrant County, Oryx sold its leases, some of which were picked up by Anadarko Petroleum. An exploration spin-off from Panhandle Eastern, Anadarko was more focused on gas than Oryx and was thus interested in Mitchell's exploration. Its tests in Tarrant and Parker Counties, however, failed to bring in enough gas to tempt the company to go further, so it too dropped plans to continue.[1] From Anadarko's perspective, the Barnett was a waste of time and money.

The most striking example of major companies' unwillingness to continue Barnett exploration in the face of an initial disappointment is Chevron's involvement in Johnson County. By 1995, Chevron had learned about George's work with the Gas Research Institute. The company had a heightened interest in going after unconventional gas. Mitchell Energy staffers were willing to share some of their Newark East information with Chevron because they, too, were curious about the potential of Johnson County, but not enough to spend money there. Not a single producing oil or gas well had ever been drilled in the county, whose mineral resources consisted of sand and gravel. In August 1997, Chevron spudded in its Mildred Atlas No. 1 in southeastern Johnson County, drilled down to the Barnett, and used a gel frack to stimulate it. The well came in for only a dismal 150 Mcf per day. Though Chevron geologists, including Kent Bowker, were reluctant to give up on the strength of only one test, Chevron not only abandoned the Johnson County exploration but sold its Barnett leases and disbanded its unconventional gas exploration team. In the end, Mitchell

Energy benefitted from Chevron's actions because it was able to hire Kent Bowker away from the major company.[2]

If focus on giant finds worked against sustained large-company interest in the Barnett Shale, economics presented a daunting barrier to smaller independents who wanted to try their luck. With income from Boonsville Bend production and its gas processing operations, Mitchell Energy could offset some expenses of Barnett experimentation. When it found gas, it had gathering line infrastructure to get gas to its processing plant at Bridgeport, and the company could sell processed gas to NGPL for a substantial contract price of $3 per Mcf by the early 1990s. Moreover, as a public company, Mitchell Energy had ready access to capital. By contrast, smaller independent competitors had none of these advantages. The cost of fracking would strain their budgets. If they found enough gas to interest a purchaser, they would probably get only $0.90 per Mcf for it, a return not likely to lure capital from investors. Worse yet, in what would seem to be the best area for Barnett production, Mitchell already held the bulk of leases and was not going to let them go. Small wonder, then, that in the decade before 1998, when George proved that less expensive slickwater fracking could suceed in the Barnett, only nine small independents tried wildcatting in the formation.[3]

After 1998, however, there was reason for smaller independents to adjust their perspective on the Barnett Shale: gas prices were finally rising as the surpluses of the 1980s and early 1990s disappeared. And though Mitchell Energy was closemouthed about its geological and technological breakthroughs, the company's tremendous expansion of exploration and infill drilling, with its skyrocketing gas production, certainly made it seem like years of experiment were paying off. There was still little shared knowledge about the formation, but there was growing reason to believe the Barnett's potential was underrated. In May 1998, Vello Kuuskraa and others published an article in the *Oil & Gas Journal*'s series on emerging US gas resources, which suggested that the shale might hold ten trillion cubic feet of recoverable gas, as opposed to a 1995 USGS estimate of 3.4 trillion cubic feet. The authors pointed out that ten trillion cubic feet of gas was the equivalent in British thermal units of "a giant 1.67 billion bbl oil field," a statement guaranteed to captivate oilmen.[4] They also argued that data from existing wells indicated that most wells drained only a limited area, perhaps only ten to thirty acres in extent, which could mean that even a small frack might yield a profitable level of production if properly drilled and produced. It would still be expensive to imitate Mitchell, but there was more temptation to try. In any event, from 1998 to the end of 2000, the number of Barnett Shale wells drilled by smaller independents jumped from 20 to 186.[5] The play took off.

Since the Barnett Shale extended over a broad section of the Fort Worth Basin, independents looking to try their luck had to figure out where best to drill on the acreage available. Common sense would lead one to "closeology"—drilling to the Barnett as near as possible to where Mitchell was bringing in good production. There were two difficulties in the way of carrying out that strategy. The first was the amount of land Mitchell Energy had already leased in Wise, Denton, and northwest Tarrant Counties. It would not be easy to find a large block of acreage near Mitchell; there was little point in picking up an amount of land that would accommodate only a few wells. Still, there were thousands and thousands of unleased acres over the Barnett, and that brought up the second problem: how little was known about drilling the area.

Effective fracking was necessary to produce much gas from a Barnett well. In Mitchell's core area (southeastern Wise, southwestern Denton, and northwest Tarrant Counties) the Viola Limestone formation lay under the lower Barnett, and under the Viola lay the Ellenburger formation, which could contain a large amount of brine. In effect, the Viola was a barrier preventing Ellenburger brine from moving into rock fractured around a Barnett wellbore. But geologists knew the Viola pinched out to the west of Mitchell's core area, beyond which the Ellenburger underlay the Barnett. Were a frack to break rock beyond the Barnett into the Ellenburger, instead of bringing in gas, one might, as Larry Brogdon put it, "bring in the ocean."[6] Or, at least, that was prevailing opinion among geologists.

Mitchell Energy, recognizing this opinion as a deterrent to competitors' leasing, did nothing to contradict it. And this was not the only hazard to fracking a good well in the Barnett. There were irregularities in the formation in the form of faults and karsts (collapsed caverns) in the rock. Send the highly pressured fluid of the frack into such irregularities, and they would siphon the fluid away from the propagating fracture, thus failing to part the rock as intended. Geologists would eventually find that using highly focused seismic technology could locate these hazards so that they could be avoided in drilling a well, but that was not common knowledge in 1998. Similarly, oilmen would learn that one could drill good Barnett wells where the Viola was not present; however, it would take more time, technology, and hands-on experience to do so. So in the early stages of the play, leasing did not take place over as wide an area as one might have expected. Independents hesitated to jump on board.

The first independents to get on the Barnett bandwagon were for the most part industry veterans already familiar with the economics of exploration in the Fort Worth Basin. They included the Four Sevens group, headed up by Dick Lowe and Hunter Enis; Chief Oil & Gas founder Trevor Rees-Jones; partners Ted Collins and George Young Jr.; and Hallwood Energy's Hollis Sullivan.

Before the Barnett Shale play took off, partners Lowe and Enis were drilling Bend Conglomerate gas wells in Wise and Jack Counties. Lowe's American Quasar had been a casualty of the industry downturn of the 1980s, leaving him strapped for cash. Enis, a geologist and TCU football star like Lowe, had returned to Fort Worth after several years of playing in the National Football League. He looked up Lowe, and they formed a partnership: Four Sevens Oil. The partners might have continued drilling Bend Conglomerate gas wells had they not asked Larry Brogdon to join them in 1998. Brogdon had a keen interest in Mitchell Energy's operations in the Barnett, piqued by George's general unwillingness to share much information about its projects. Working out in the field for Lowe and Enis, he did what he could to find out as much as possible about what George was doing. "I kept listening and scouting,"

he said. "I'd go out to wells and check on them . . . and jump fences and do whatever I had to do."[7] Believing that Mitchell had uncovered something really big, Brogdon began to push Lowe and Enis to try drilling the Barnett, and in 2000, the partners agreed to bring Brogdon on board for that objective, settling the deal with a handshake over lunch at the Fort Worth Club.[8]

At the time Brogdon joined Lowe and Enis in partnership, he was also president of the board of directors of the Oil Information Library of Fort Worth, which had fallen on hard times. Not only was their building badly damaged by the tornado that roared through downtown Fort Worth in the spring of 2000, but many companies were relocating offices and personnel to Houston, thinning the ranks of library members. Realizing that he was only one of many oil professionals with a growing interest in the Barnett Shale—and, coincidentally, what Mitchell Energy had learned about it—Brogdon decided that the Oil Information Library would hold a Barnett Shale Symposium. It would invite geoscientists to share what they thought about the Barnett, thus furthering knowledge, while registration fees would raise money for the library. The Symposium, held on September 28, 2000, at the Petroleum Club of Fort Worth, was more successful than anyone could have foreseen: so many oilmen showed up that the city fire marshall eventually had to turn away latecomers. Curiosity about George's findings remained unsatisfied, since George's geoscientists were conspicuously absent. The Symposium stimulated more interest in the Barnett, perhaps encouraging more industry participation in the play. But it left the critical question unanswered: how much of the Barnett Shale would be productive? If George already had most of the area likely to have good wells, there was not much hope for newcomers looking to make money drilling the Barnett.[9]

Unlike the majority of local oilmen, Brogdon did not believe that one could only frack the Barnett effectively in the part of it lying over the Viola Limestone. Instead, he thought the Barnett could be productive over a tremendous area in the counties around Fort Worth. Watching Mitchell Energy lease outside its core area in places like Parker and Johnson Counties, he began to think that even George did not subscribe to the idea that the Viola was essential to effective fracking. The chance to prove his hunch was right came with a well George drilled near the Wise County hamlet of Boyd; as he put it, "I knew that well was west of the Viola pinch out." Following what Mitchell Energy's service contractors were doing, he was able to gauge when the well might or might not come in—as it happened, on a rainy Sunday. He had taken his family to church, and having dropped them off back at home, went out to the well:

> I still had on a suit of clothes. And it was wet. And I went to that lease. The gate . . . was open. . . . I drove in there, and it was muddy. And I was out there in my suit pants and Sunday shoes on, wading out there. . . . They had it [the well site] all locked up, but I could hear it. And I knew the sound, I could tell it wasn't dumping a tremendous amount of water. . . . This well was good, and it was beyond the pinch out. And at this point I knew that this was gonna be huge.[10]

In fact, by the time Brogdon waded out to the Mitchell well, Four Sevens was already embarked on an aggressive leasing program in northwest Tarrant County around the town of Haslet and in Parker County. In the latter, they were able to go into leasing by joining forces with Dallas independent Denbury Resources. Though in some areas they were competing for acreage with Mitchell Energy, personal connections with farmers and ranchers in Parker County helped them pick up some 15,000 acres there and additional land in Johnson and Hood Counties. For the most part, what they paid for leases ranged between $100 and $150 bonus an acre, a rock-bottom price compared to what would happen later on in the boom. As Brogdon reflected, "Eventually it got crazy, $27,000 [bonus an acre], I mean, just nuts."[11] But even with more modest leasing costs, Four Sevens would need more capital to

keep growing. Luckily, through a friend, former Union Pacific Resources vice president Marty Searcy, they were able to find an important new investor.

A master of the art of putting entrepreneurs in touch with investors, Marty Searcy directed acquisitions and divestitures for Union Pacific Resources and had a wide circle of industry contacts. After Union Pacific sold out to Anadarko in 2000, Searcy began to work with Johnny and Bud Vinson, who had a five-thousand-acre ranch in southeastern Wise County, right in Mitchell Energy's core area. Inspired by George's recent success, the Vinsons wanted to drill Barnett wells on their ranch and were looking for investors. Searcy found them a partner in Salt Lake City-based Sinclair Oil Corporation, headed up by Ross Matthews. The partners' company Threshold Development successfully drilled ten Barnett wells between 2001 and 2003. In 2003, the Vinsons decided to sell out to a Denver company, Antero Resource Corporation, whose CEO Paul Rady was ready to pay top-dollar prices for leases well-positioned in the emerging Barnett play. But if the Vinsons were ready to cash out, Matthews was not. Accordingly, Searcy introduced him to the partners in Four Sevens, and the Sinclair group entered into a fifty-fifty partnership with them. As Larry Brogdon reflected, "It was a great relationship." Four Sevens embarked on an ambitious leasing program in northern Tarrant County, leasing "ranchettes," properties of four to six acres, and eventually within the city limits of Haslet, their first experience in urban drilling. By 2004, their acreage and production led to an offer of purchase by rapidly growing XTO Energy, and Four Sevens sold out to them for $155 million. In fact, Lowe was ready to bargain for a higher price for the company's twenty-six thousand acres, but in negotiating, when XTO offered $155 million, Lowe recalled, "Before I could say no, Hunter [Enis] said, 'We'll take it.'"[12]

Selling out did not mean retreating from the Barnett play. Instead, Lowe became focused on the Barnett, reading everything he could find about it and studying the play's general pattern of development by mapping existing production. To judge from wells in place, at the beginning of 2005, the play seemed to be moving south through Tarrant and

Johnson Counties. Maps in hand, he told Enis, "'Hunter, this is the biggest cinch that ever happened. . . . We need to go down on the south side of Fort Worth and start leasing' . . . so we went down there and started leasing like crazy with our partner Sinclair." Beginning in 2005, Four Sevens picked up thirty-nine thousand acres, largely in southern and eastern Tarrant County and the Trinity River Valley, as well as substantial acreage in northern Johnson County, heedless of whether or not the Viola formation was present under the Barnett. Much of what they picked up was in small tracts of five to ten acres, but they outbid XTO to lease Spinks Airport in south Fort Worth for $4,000 an acre, double what then prevailed in Tarrant County leasing, and 27.5 percent royalty. By March 2006, Four Sevens had eleven wells producing on their Tarrant County acreage, some good for as much as seven to nine million cubic feet of gas per day, and bigger independents, especially Chesapeake Energy and XTO, were interested in buying what Four Sevens owned. In June, Chesapeake bought their leases, producing some 37 million cubic feet of gas per day, for $845 million—a price Lowe insisted upon because, taken with their previous sale to XTO, Four Sevens could say they had sold assets for a total of a billion dollars in less than two years![13]

Several weeks before Four Sevens made its dazzling sale to Chesapeake, another independent, Chief Oil & Gas, made an even more spectacular sale of Barnett properties to Devon for $2.15 billion. The sale was the culmination of over a dozen years of Fort Worth Basin exploration by Dallas independent Trevor Rees-Jones.

Rees-Jones chose an unpromising time, 1984, to go from being an oil and gas bankruptcy attorney to being an independent oilman; only eighteen months later, the domestic petroleum industry hit the lowest point in its history since the Depression. He survived the downturn, and in 1994 decided to focus his exploration on the Fort Worth Basin. He set up Chief Oil & Gas, naming his company after his Labrador Retriever, and started drilling gas wells in the Bend Conglomerate, picking up southern Wise County leases outside Mitchell Energy's extensive inventory. As he drilled what George passed up, he seldom sold his gas for as much as $2 per Mcf, but what he was doing did not entail much risk. As he told one interviewer, "I was going to pick the last meat off the bone." Working in a small way in the heart of George's turf, however, it was not long before he became intrigued by the company's Barnett tests and, in 1997, tried his own luck with a Barnett well. It was not a spectacular success, but Rees-Jones did not give up. Instead, tracking what George was doing, he started leasing acreage in Denton and northern Tarrant Counties, and he continued drilling Barnett wells. By 2002, Chief had six thousand acres in Denton County, ten thousand in northwest Tarrant County, and forty-two Barnett wells with a production of 20 Mcf per day.[14] To keep growing, however, Chief needed more land.

Just as it seemed Rees-Jones would have to move headlong into town lot leasing in the Fort Worth suburbs, a lucky alternative appeared in the form of a project that two other independents, Ted Collins Jr. and George M. Young Jr., were putting together. Both men came from families long active in the Fort Worth independent community. Collins's father, C. O. Collins, had been active in exploration in the Permian Basin and East Texas during the twenties and thirties; based in Midland since the 1960s, Ted Collins had been one of Dick Lowe's board members in American Quasar. As a partner in Marshall and Young, George Young's grandfather had brought in oil production on the Cook Ranch in the 1920s, and Young himself had been operations manager of his family's business. Young decided he wanted to go out on his own and enter the Barnett action heating up in 2001, but he needed a partner. He found Ted Collins. The two men agreed to become partners at a Rangers baseball game, writing out their agreement on the back of a napkin. Collins would raise money, Young would round up leases, and they would share the partnership on a fifty-fifty basis.[15]

As George Young started looking for land to lease, three types of online resources made his work much easier, and they represented information innovation in the Barnett play. Turning to the Tarrant Appraisal District website, Young could pull up tax rolls showing not only current land ownership but also lists of previous owners, often with dates and deed references. He recalled, "I was able to do preliminary land work without ever setting foot in the courthouse." Another online resource, Courthouse Direct, allowed him to look at deeds filed after 1977; sometimes this resource would even give the deed book and page numbers for a given document, cutting the time he had to spend in his searches to a minimum. The third resource, DrillingInfo, let him see up-to-date data on individual wells' locations, completion data, and production, the sort of information one could have gotten working through a host of Railroad Commission records. These online resources meant a tremendous saving of time, and within a year Young was able to lease over three thousand acres, some in Denton County near Argyle, and some from ranchers and landowners he knew in northern Tarrant County, where Young's personal friendships counted heavily. With leases assembled, it was time to drill some wells. To do so, Collins and Young needed more capital and someone ready to take on the role of operator.[16]

Collins and Young found exactly the help they needed from Rees-Jones. His Chief Oil & Gas not only had substantial financing from Dallas bankers but also a top-notch geoscientific staff, including many geologists and engineers formerly with Oryx. Moreover, at a time when idle drilling rigs were not easy to find, Chief already had two rigs drilling Barnett wells. For their part, Collins and Young had a highly desirable amount of land under lease. Young came up with a brilliant strategy to get more: leasing the Alliance cargo airport north of Fort Worth and the industrial development around it, an area spanning some seventeen thousand acres. Once again, Young's personal friendships were key to success. The Alliance development was the brainchild of H. Ross Perot Jr., and his Hillwood Development managed its real estate. As it happened, Young had known Perot from college days, and a close friend of Young's managed Hillwood. It took the better part of a year to work out a partnership between Chief and Hillwood, but the association was formalized in August 2003. Chief drilled the first four wells vertically; then Rees-Jones decided to drill the rest horizontally, thereby keeping airport disruption to a minimum. Wellbores thus snaked under the airport and surrounding industrial complex. When these horizontal wells were successful, Chief started using horizontal wells exclusively, being one of the first smaller operators to do so. Drawing inspiration from offshore technology, Chief also began to drill multiple wells from one pad site, a technique that would become essential in subsequent urban drilling.[17]

Chief's Alliance project was a model for oil and gas development in an urban industrial environment. It was also a tremendous success. For Collins and Young, the Alliance participation was, as Young put it, "the crown jewel of all the stuff we did with Chief." Young himself acted as liaison and point man on-site between Chief and Hillwood, officing in a double-wide trailer on the Alliance runway. But as well after well came in, the problem of market outlet became critical. Rees-Jones decided he would put in his own pipeline infrastructure and organized Eagle Mountain Pipeline Company, a Chief subsidiary pipeline, and processing operations would eventually grow beyond Denton and Tarrant Counties into Parker County. By 2005, production from Chief's 328 wells made the company one of the largest gas producers in the Barnett play and an attractive target for acquisition. Devon bought out Chief's wells and a total of 169,000 acres for $2.15 billion, while CrossTex Energy bought the pipeline and processing operations for $480 million. Disappointed with Devon's offer to buy them out at the same time, Collins and Young would eventually sell a large part of their interests to the Southern California Public Power Authority (SCPPA). In any event, in partnership with Collins and Young, Rees-Jones did much better than "pick the last meat off the bone."[18]

Like the independents of Four Sevens and Chief, third-generation independent Hollis Sullivan had been active for some years before entering the Barnett action. Forming his first company in 1990, he discovered the Pipe Dream field in Concho County, not a large find but one that helped him get through industry doldrums of the 1990s. Until the end of the decade he focused on oil, but by 2000 looking for gas in the Barnett had growing appeal. Gas prices were higher than they had been, and the Barnett was emerging as a relatively low-risk play—dry holes were not usual. Most of all, the new slickwater frack technology made wells a whole lot more affordable than they were earlier. Like most other small independents who entered the play early, Sullivan looked to find a place to drill close to Mitchell's turf. Operating as Hallwood Energy, he put down his first Barnett well in 2000 in southwest Denton County, a few miles west of the small town of Ponder. The well came in with unspectacular but satisfactory gas production, encouraging Sullivan to move south and pick up leases in northwest Tarrant County. In the area he was entering, the Viola did not underlie the Barnett, but drilling horizontally could avoid fracking into the Ellenburger and having water problems. He drilled his first horizontal well in 2003, and while it was more expensive than a vertical, it was a success.[19] In fact, operators like Sullivan and Rees-Jones learned that, while horizontal wells cost more than vertical ones, their cost was more than offset by better production.

Sullivan's success encouraged him to expand in Tarrant County, so it wasn't long before he was leasing smaller properties, tracts one to five acres in size. His landmen made a point of trying to explain to landowners what leasing would mean and what they could expect to see once operations were underway. A lot of explaining was necessary. As Sullivan recalled,

> I was asked one time, is there any possibility that the drill bit is going to come up in my swimming pool, you know, stuff like that. We actually had a landowner one time ask us, they thought the gas was [transported] when . . . a truck came up and filled up . . . they asked how much it would cost to truck that gas. . . . [20]

Some landowners made a point, even in 2002–2003, of bargaining with rival landmen for the best deal. Thus, one lady in Denton County, with a sizeable property of close to two hundred acres, told company representatives that, in addition to an up-front bonus, she wanted various property improvements: "She wanted a jogging track built, she wanted a new swing set for her kids, I mean, it was unbelievable the laundry list that she had."[21] Sullivan backed away from leasing her property.[22]

As the play moved south toward Fort Worth, suburban town lot leasing, "roof top leasing" as it was dubbed, picked up in 2003–2004. Sullivan reflected, "Everybody avoided the urban drilling as long as they could, but it got to the point where there wasn't anything left," or so it seemed. For enough land to make drilling worthwhile, it became necessary to work with dozens of landowners in neighborhood meetings. One of the first such meetings Hallwood would hold was in the Fort Worth suburban community of Haslet. Company representatives held a Saturday afternoon barbecue, told landowners what they wanted to do, and came armed with leases ready to sign on each property. By 2005, Hallwood had some sixteen thousand acres in Denton and northern Tarrant County, and fifty producing wells—twenty-five vertical and twenty-five horizontal. Sullivan decided to sell out to Encana since, by this time, he had shifted his focus to northern Johnson County and the bedroom communities south of Fort Worth.[23]

If one looked at a map of Barnett wells in 2003, one could guess that the trend of the play lay right through downtown Fort Worth and south into Johnson County. Although Mitchell Energy had picked up Johnson County leases before selling out to Devon, others hesitated to enter the county because much of it lay west of where the Viola separated the Barnett from the briny Ellenburger. But when Devon drilled its horizontal well Veale Ranch No. 1 in 2002, far west of the Viola's limit in Tarrant County, it demonstrated what Sullivan himself would find the following year: horizontal drilling could bring in good wells even in the absence of the Viola. That breakthrough opened up the possibility that a vast area of the Fort Worth Basin could have lucrative Barnett production. From a wildcatter's perspective, Johnson and Parker Counties were the next most promising frontier. Sullivan picked up thousands of acres of leases in northern Johnson County, once again leasing small tracts. By mid-2004, Hallwood Energy's Johnson County gas wells were producing over twenty million cubic feet of gas per day. Its main problem, one shared with other area independents, was lack

of pipelines to carry gas to market. Sullivan resolved that problem in December 2004 by selling what he had in the county to Chesapeake for $277 million.[24]

Hollis Sullivan's sale of Barnett properties to Chesapeake was part of a trend well underway by the end of 2004: the growing participation of larger independent oil and gas companies in the Barnett Shale play. Some, like Burlington Resources and EOG (a spin-off from Enron), entered the play before or during Devon's buyout of Mitchell Energy but hesitated to make it a major focus. By the beginning of 2005, with natural gas prices moving above $5 per Mcf, these larger independents were ready to ramp up operations. EOG, for example, had over 400,000 acres leased in the Barnett, much of it in Johnson County; it was expanding into Parker, Jack, Hood, Hill, and Erath Counties. It planned to put down as many as ninety Barnett wells in 2005. Others like XTO and Chesapeake shifted emphasis to the Barnett and launched aggressive campaigns to dominate the play. As Young perceptively summed up this turn in Barnett action, "We did all the land [leasing], we did the footwork . . . so we kind of all set the table for them, and now they're at the banquet"—having, of course, made the fortunes of the smaller independents who sold out to them.[25] Seen in historical perspective, however, smaller companies opening a play and selling out to larger players was a phenomenon born with the American petroleum industry. The Barnett saw it happen once again.

In terms of meteoric growth, few companies rivaled Fort Worth-based XTO. The company had its origins as Cross Timbers Oil, founded by three former Southland Royalty employees: Jon Brumley, Bob R. Simpson, and Steffen Palko. In 1996, Brumley left to form his own company, and in 2001, Cross Timbers changed its name to XTO, with Simpson as chairman and chief executive officer. From there on, Simpson led the firm in a dizzying pace of acquisitions and expansion that went beyond natural gas into Fort Worth real estate, buying office buildings to house its burgeoning ranks of employees. In March 2004, XTO bought Four Sevens's Barnett properties. Nine months later it agreed to purchase Antero Resources's Barnett holdings in Tarrant,

Johnson, and Parker Counties, wells producing some sixty million cubic feet of gas per day, for $685 million. The purchase made XTO one of the top Barnett gas producers, second only to Devon. In 2006, it bought Peak Energy, a Colorado firm with acreage in the Barnett; in 2007, it spent $2.5 billion to buy Dominion Resources; and in 2008, it bought Hunt Petroleum for $4.19 billion.[26] By that time, XTO was picking up gas reserves far beyond the Barnett.

Compared to XTO's headlong plunge into purchasing Barnett properties, the strategy pursued by Chesapeake Energy, at least between 2002 and 2005, was relatively cautious. After that, the Oklahoma City-based firm would rival its counterpart in a rush to grow, both in the Barnett and in other shale basins. Chesapeake had bought a share in Hallwood Energy's Johnson County exploration in 2002, following this with the 2004 purchase from Hollis Sullivan. With that acreage in hand, it began an ambitious drilling campaign in the county, hoping to double production from this source to fifty million cubic feet of gas per day.[27] Doing so, however, would mean going right into the challenge of urban drilling. When it purchased Four Sevens's southern Tarrant County acreage in 2006, and its downtown acreage the following year, Chesapeake committed itself to a leading position in downtown drilling and a host of problems that smaller and warier independents tried to avoid. But such problems were easy to overlook in mounting boomtime excitement.

Notwithstanding ambitious rivals like XTO and Chesapeake ready to challenge its dominance in the play, Devon kept its leadership in Barnett gas leases, wells, and production, essentially maintaining the lead it got by buying Mitchell. Apart from the amount of acreage Devon acquired, the geoscientists it inherited from Mitchell gave it effective first-mover advantage in understanding the shale and using technology to exploit it. It lost some of that advantage, however, when it decided to close its Fort Worth office to concentrate operations in Oklahoma City. Seven of its leading engineers—the Devon Seven, as local Fort Worth independents would call them—decided not to move to Oklahoma. Instead, they went independent and decided to hire out their expertise to Devon's smaller shale competitors, thus letting them share the knowledge Mitchell spent so many years and dollars to achieve. Nick Steinsberger, for example, worked for Charles Moncrief and Four Sevens, agreeing to help the latter as a completions expert. Dick Lowe asked Steinsberger what it would take to hire him. According to Lowe, Steinsberger said, "'I'll do that if you buy me some health insurance.' I said, 'You're on!'" Particularly with respect to the technology of using horizontal drilling in the Barnett, the hired talent of the Devon Seven was invaluable to independents without their own corps of engineers.[28]

Mitchell Energy had experimented with horizontal drilling, but since it was significantly more expensive than vertical wellbores, Mitchell had not gone far with it. By 2002, however, prospectors had learned that horizontal drilling expense was more than offset by far greater well production. Like smaller independents, Devon engineers learned that using it could sidestep fracking problems outside the area of the Viola, thus allowing them to expand far outside their core area. Plunging into horizontal drilling, Devon advanced into acreage Mitchell had leased in Johnson County as well as in eastern Parker County. Not that it abandoned its core area; it just picked up 90,000 more acres in Wise and Denton Counties in 2002. That additional acreage would allow putting down some 1,300 wells on fifty-five-acre spacing, but, by 2002, it was clear that wells on twenty-seven-acre spacing would not mean excessive drainage because gas molecules in the rock were tightly locked. In any event, the way was clear for Devon to pursue a mammoth drilling campaign, which it did. Within two years Devon had drilled its thousandth horizontal well. It had, moreover, refracked hundreds of older wells Mitchell had drilled, often bringing them back to original production or better. With its wells producing 570 million cubic feet of gas a day, by mid-2005 Devon was not only the lead producer in the Barnett but also the largest gas producer in Texas and third largest in the US. And the company was ready, as with its purchase of Chief's properties, to expand even more.[29]

At this point, one could ask which fence the major companies were sitting on while independents were finding and producing billions of cubic feet of natural gas. Part of the explanation for their inactivity lies in the general shift of major company exploration strategy after 1980, a shift which took them away from domestic onshore projects to apparently more promising arenas offshore and overseas. That shift reflected the majors' focus on elephant hunting, looking for the "really big," both in terms of additions to reserves and the resources necessary to exploit them. Smaller projects simply were not worth taking up the time of their large exploration and production teams. Thus, when the Barnett play seemed confined to a small part of three Texas counties and only promised potential reserves of fifty billion cubic feet of original gas in place per square mile, it was beneath their notice. But by 2005, when it was clear that the Barnett offered possibilities within nineteen or more Texas counties, with close to three times the estimated original gas in place and wells whose production looked to be good for as much as thirty years, the play might seem more alluring. Unfortunately for the majors, by that time it was harder to put together the kind of enormous blocks of acreage they wanted and needed.

Notwithstanding disadvantages to entering the play late, three majors—ConocoPhillips, Royal Dutch Shell, and ExxonMobil—decided to buy in. In December 2005, ConocoPhillips bought Burlington Resources, which had itself bought out independent Republic Energy of Dallas, an early player in the Barnett action, the previous August. Also in August, Shell told journalists it was "rediscovering" unconventional gas and bought acreage in five counties, primarily in Parker County. At the same time, it was negotiating to purchase Chief's properties and to lease DFW Airport. Had negotiations succeeded, the company would be on its way toward enough critical mass to justify staying in the play. As it was, Devon bought out Chief, and Chesapeake would drill under the airport. Shell drilled thirty wells on what it had, but failure to lease much more land made its Barnett participation disappointing, and it would sell out to Pioneer Natural Resources in 2007.[30]

Though ExxonMobil's corporate headquarters were in Irving, the oil giant did not enter the action taking place virtually in its backyard, as one journalist put it, until approached by a small Fort Worth independent, Rick Harding. The Harding Company had picked up leases in the Metroplex along an old Mobil pipeline going from Keller to Corsicana. If Harding could persuade ExxonMobil to let the line be used to ship his gas, he could solve the problem so common among Barnett independents: the lack of pipeline infrastructure. Talks began in 2004, and at the end of 2006, Harding and ExxonMobil created a partnership between Harding Company affiliate, Cinco County Barnett Shale, and an ExxonMobil subsidiary, Metroplex Barnett Shale, the partnership to be called DDJET Ltd. The name reflected the counties the partnership intended action in—Dallas, Denton, Johnson, Ellis, and Tarrant Counties. Cinco County would pick up leases and permits; Metroplex would operate drilling and production, and gas produced would go into the old Mobil line. The arrangement allowed ExxonMobil to try its luck in the play while keeping a low profile and gave Harding a way to drill wells and ship out the gas they produced.[31]

By the end of 2006, the Barnett Shale action was increasingly a game for big players. Costs of leases and drilling escalated dramatically, requiring enormous amounts of capital to stay in the game. Smaller independents did not abandon the Barnett, but as the cost of leases in the most promising areas skyrocketed, many who sold out to bigger players for handsome profits either cut back their involvement or ventured into relatively untried parts of the shale. So as drilling rigs moved into city neighborhoods, college campuses, and downtown, those with deeper pockets came to dominate the action, and of that group, Chesapeake was most aggressive at pushing forward inside city limits. But moving into town would make the problems of urban drilling noticeable—and more controversial.

CHAPTER 4

Into the Boom

By 2006 it was clear that the Barnett Shale play had produced a massive regional boom along with gigantic quantities of natural gas. The boom would bring in thousands of workers, both in oil field work and in supplying everything the industry and workers needed. But the communities and urban neighborhoods seeing workers and rigs come into town were far from prepared to cope with problems generated by an industrial invasion. True, the boom meant an incredible economic windfall. But how were urban industry operations to be adjusted to guarantee neighborhood comfort and, more importantly, safety? When city governments and communities struggled to address new and pressing problems, there were no convenient models for action. As for industry members, they, too, were on unfamiliar turf: usual operating procedures out in the country were not necessarily suitable within city limits. The breakneck pace of the boom, however, left scant time to reflect on new problems.

From the beginning of 2006 to the end of 2007, Barnett Shale action generated spectacular growth in metropolitan Fort Worth and out into adjoining Denton, Johnson, and Parker Counties, where leasing had been lively for months. But drilling rigs also followed landmen south and west into Bosque, Ellis, Erath, Hamilton, Hill, Hood, and Palo Pinto Counties. That no one knew exactly what the play's limits would be merely spurred wildcatters on. By the end of 2007, some daring prospectors were even leasing in Lampasas County, a hundred miles south of the play's epicenter around Fort Worth.[1] As with any boom, it was easy to believe that fortunes had nowhere to go but up.

As the play expanded geographically, however, wildcatters learned that drilling into the Barnett was far from the same everywhere. Moving west into Parker County, drillers were more likely to encounter underground rock anomalies such as unexpected faults or karsts. Unless located by seismic work and avoided by the drill, these anomalies could derail effective fracking. The farther west one went in Parker County and beyond, the more likely one would encounter gas containing more liquids or even oil, especially where the Barnett pinched out in Erath and Palo Pinto Counties. Wildcatting in the Barnett to the north in Montague County could yield oil, as EOG found in 2008. Variable conditions within the formation, then, challenged geoscientists looking for the best ways to find and produce the gas and oil the Barnett contained.[2]

Ironically, in terms of the geographical extension of the play, Dallas County was not very tempting to Barnett prospectors. Between Dallas and Tarrant Counties lay an underground geologic structure, the Ouachita Thrust, which seemed to mark the play's eastern limit, or at least where the Barnett deepened, thickened, and might be far less productive. In 1997, Dallas wildcatter Sanford Dvorin, who had been following Mitchell Energy's Barnett experiments, brought in a gas well in Coppell, but its production was too small and gas prices were too low to spark more Dallas County wildcatting. A decade later, geoscientists were at best skeptical of Dallas County potential, though leasing advanced as far as Irving, and Chesapeake Energy planned drilling under DFW Airport. What was obvious by 2007 was that a huge sweet spot of the Barnett Shale lay right underneath metropolitan Fort Worth, leading to the quip that the Barnett Shale was "God's joke on Dallas." Had he lived to see it, inveterate Fort Worth booster Amon Carter would have laughed longer and louder than anyone.[3]

However uncertain the Barnett Shale's productive limits were, it was absolutely clear by 2006 that the play was having a profound effect on the North Texas economy, especially in Denton, Tarrant, and Wise

Counties. Some measure of the impact of gas drilling may be seen in a study by former University of North Texas (UNT) economists Bernard L. Weinstein and Terry L. Clower on the effect of spending in 2003 by Devon Energy, the play's largest player. Perhaps with an eye to public relations, Devon commissioned the scholars to look at drilling's economic contributions to the three counties where it was then most active—Denton, Tarrant, and Wise.[4]

Weinstein and Clower looked at a wide range of impact areas, including Devon's direct spending on exploration and production; the impact of its operations on the contractors and companies with which it did business; spending generated by wage earners; income to local governments from royalties, bonuses, and property taxes; returns to landowners from leasing and royalties; and overall population growth. Despite looking at only one company for only one year, their findings were stunning. They found Devon directly responsible for over $475 million in economic activity, including $96 million in salaries and wages to over two thousand employees. Wealth derived from royalties, rents, and corporate profits was close to $109 million. State and local government taxes paid by Devon amounted to $22.5 million. Mineral-based property taxes in Northwest Independent School District, for example, brought in more than $36 million, though not from Devon alone.[5] Overall, in terms of wealth, the scholars concluded that Devon's operations generated $611 million.[6]

Jobs and opportunities created by soaring wealth acted as a population magnet. Though large parts of Denton and Wise Counties were still rural, between 2000 and the end of 2003, population in those counties rose 17 percent and 11 percent respectively. Notwithstanding national economic slowdown, employment in these counties rose 4 percent and 16 percent, respectively, during the same period. During the same time, Dallas County, outside the play, lost 4 percent of its jobs.

In a follow-up study of 2005, which Weinstein and Clower completed in 2006, the economists found even more impressive gains. By this time Devon was active in nine counties. The total impact of its operations amounted to $1.36 billion, supporting over 8,800 jobs. In seven of the nine counties, the percentage of population growth since 2000 was in the double digits; in five years, the population of largely rural Johnson County rose 13.4 percent and Hood County 16.9 percent. As the population grew, so did personal income levels, rising far more in the nine counties between 2000 and 2004 than in either Dallas County or the Dallas-Fort Worth Standard Metropolitan area as a whole.[7] In short, workers were pouring in, finding jobs readily, and, like long-term residents, spending money. A boom was well under way.

Drilling into the Barnett required thousands of workers in a myriad of occupations. Once company geologists decided where to drill and landmen went out and leased land, contractors had to prepare the well site, which meant clearing where the rig and equipment would go, clearing roads to the well sites, and digging and lining mud and water pits. When the rig was in place, truckers would bring in casing, drill pipe, and other materials, and electricians would make connections to bring power to the site. Apart from the roughnecks on the drilling crew, rig operation also called for service workers in cementing, fracking, mud logging, pipeline construction, and salt or frack water collection and disposal. All these workers needed places to eat, sleep, and have fun—although the physically demanding nature and long hours of their work could limit capacity for the last. As in all booms, housing was in short supply, particularly in rural areas, so employers brought in mobile homes, housing anywhere from four to twelve workers per unit, depending on crew size and arrangement of work shifts; motel space typically had long since been booked up. Thus, most married workers left families elsewhere. Few tried to commute from homes in the Metroplex: the demands of work did not mix well with commuting.

Once drilling was underway, work usually proceeded around the clock. Most Barnett drilling contractors had workers on twelve-hour shifts: twelve hours on, twelve hours off, seven days a week, with seven days on and seven days off. On their week off, roughnecks were free to go back to their families if they chose. Other workers followed much more varied schedules. A mud-logger, for example, who monitored what the drill bit brought up from underground, often worked twenty-four hours in a trailer on site, followed by twenty-four hours off. Truckers might work more conventional hours, but could expect to be on call even during off times. And work continued regardless of heat, cold, or rain. With good reason, country boys used to outdoor work found it easiest to adjust to oil field routines.[8]

Handsome paychecks did a lot to offset long hours and hard work. As Denny Smith of Nabors Industries commented, "The lowest level guys are now making $40,000 to $50,000 a year." A floor hand on a rig could bring in between $55,000 and $61,000 a year, while a driller could make $80,000 or more. Nor did it take long to advance to a higher position. Tool pusher Javier Naranjo reflected, "Now you can work a year and a half and move up to driller." In addition to hefty hourly wages, employers attempted to recruit and keep workers by offering not only full benefits, including health insurance, but also a great variety of bonuses—for taking the job, for staying as long as six months, for pursuing a healthy lifestyle (the so-called "living well bonus"). Bonuses, however, did not always keep workers around: some hands signed up, took a bonus, and left after several weeks to get a hiring bonus with another firm. But especially for workers who had had low-paying jobs in a rural environment—the convenience store manager, the cable TV installer, the local small-town government employee—taking oil field work meant jumping from a single-digit to a double-digit hourly wage. Suddenly the new bass boat and the new truck to haul it to the lake were affordable.[9]

Even with economic incentives, employers scrambled to find workers willing to take industry jobs. In part, the shortage of workers was the heritage of industry downtimes. Many of those who lost industry jobs in the busts of the 1980s and 1990s never returned to the oil and gas workforce; job security could mean more than a large paycheck. The image of industry work as work that was hard, dirty, and dangerous, which was not entirely misleading, also worked against recruiting new workers. Employers had best luck recruiting workers from the rural communities with limited job opportunities, like many communities in Parker and Johnson Counties. The operations manager for one

Parker County oil field service firm, for example, hired two-thirds of his workforce from county residents. David Watts of Midland-based Quality Logging observed, "We started hiring people up there [Johnson County] . . . you know, we'd hire truck drivers or just whatever we could find and train them. . . . It was real easy to train people." In fact, Quality Logging hired a waitress in a local restaurant to be the first secretary for its Johnson County office: "That's all we could find." But even though country folk came forward to take jobs, the boom-time shortage of workers got worse rather than better in 2006 and 2007. In August 2007, BJ Services brought in seventy workers from Denver for its Barnett Shale operations. As District Manager Mike Ware said, "We need to grow faster than we can hire them locally."[10]

Companies and contractors tried to respond proactively to the acute labor shortage. Since many jobs required computer skills as well as technical experience, they encouraged regional community colleges to launch company-sponsored oil field training programs, like Tarrant County College's "Barnett Shale Roustabout 101." In the fall semester of 2008, North Central Texas College began offering a two-year technical degree including courses in computer applications, production methods, and industrial safety, as well as a thirty-hour certificate program for workers already in the field. Since geologists and engineers were also in short supply, and many nearing retirement age, industry leaders also pressured universities to do more to encourage students to pursue the geosciences. TCU, for example, set up its Energy Institute, headed by Ken Morgan, in 2006. As promising as these programs were in terms of training industry participants for the future, however, by the spring of 2007, there was urgent immediate need for people to fill industry jobs. Thus in May 2007, companies and contractors sponsored a daylong Barnett Shale Expo at the Fort Worth Convention Center. Thousands of visitors toured exhibits, and many firms used the event for worker recruitment. The success of the expo encouraged employers to hold a Barnett Shale Job Fair at the Radisson Hotel in south Fort Worth on December 1. Twenty-two companies advertised openings and interviewed job seekers at the event, which was attended by over five thousand people. Even heroic special events, however, did not satisfy an apparently insatiable industry demand for workers. A Baker Hughes manager observed, "The [Barnett] production has been so tremendous that it's really outstretched the number of experienced workers available. . . . The labor pool just isn't keeping up with the work volume."[11]

Workers were in short supply, but rigs to drill the Barnett gas bonanza were even scarcer. This, too, was the legacy of industry hard times in the 1980s, during which the rig count plummeted, many contractors went out of business, and many a rig was sold for scrap. By September 2005, operators struggled to find rigs available for projects, often waiting months to drill. As Dick Lowe described the situation, "If you're an established producer with relationships with the driller and some contracts, you'll be able to get equipment. . . . If not, you'll have a rough time." Some contractors made operators commit to renting a rig for a year or longer as a condition of doing business. Not surprisingly, the cost of renting a rig skyrocketed, from roughly $6,000 a day in 2000 to between $12,000 and $13,000 a day in 2005. By April 2006, the cost of the average Barnett Shale well hovered around $1.2 million, roughly four times what it had been at the beginning of the decade. Encana, the second largest North American natural gas producer, actually responded by cutting back its drilling program, but Encana was in the minority. As natural gas prices reached levels no one had ever seen before—above $15/Mcf in the 2005 aftermath of Hurricanes Katrina and Rita, down to $7.69/Mcf in October 2006—still high—they kept the Barnett action going despite tremendous escalation of costs.[12]

By far the most startling price increases were in leases, especially as the Barnett action moved into town. Since the boom saw operators pay spectacular bonuses and agree to equally spectacular royalty percentages—beyond anything in petroleum industry history—the story of what happened in leasing is worth some background and explanation.

One of the peculiarities of the petroleum industry as it developed in the United States resulted from landowners' ability to own the minerals beneath the surface of their land. If they wish, they can sell these minerals (the mineral rights) to someone else, but unless landowners part with these rights in one way or another, they retain their ownership. An oilman wishing to put down a well must obtain access to minerals from their owner. He can, of course, buy a parcel of land, including whatever lies under its surface, but if his test is a dry hole, he is left with land he doesn't need. It is much more efficient for him to lease the right to obtain the minerals under the surface. Customarily, the oilman pays the mineral owner a bonus upfront to do so. The size of the bonus usually varies with the likelihood that there will be an attractive amount of petroleum recoverable, but before the Barnett action, a bonus of over $1,000 per acre would have been seen as a handsome one indeed, reflecting near certainty of large and long-lasting production.

If the oilman leasing the right to obtain minerals has a successful well, the terms of his lease with the mineral owner traditionally includes the mineral owner's right to a percentage of production, a royalty. For much of the twentieth century, the customary royalty was 12.5 percent, paid from the sale of produced petroleum rather than actual barrels of oil or cubic feet of natural gas. The lease ordinarily is for a specific number of years, during which time the oilman has the option to drill; if he does not drill and produce oil or gas within the specified time period, he typically loses the lease. During the first half of the twentieth century, leases were commonly for five to ten years. If the oilman gets production, he can continue to extract oil and gas as long as production holds up. These and other features of a lease are, of course, subject to negotiation between mineral owner and the person leasing. But the basic feature of the lease is that which gives the lessee the right to minerals.[13] Its specific terms are what lessor and lessee agree to.

What happens when an owner of the land surface does not own the minerals under his land, a common situation in areas such as the Texas Permian Basin, where oil and gas have been produced for a long time? In Texas law, the person who owns or has leased mineral rights has the right to get the minerals. If an oilman wants to drill a well on land whose owner doesn't own the minerals underneath it, he has the right to do what is necessary to drill and produce minerals—to clear a rig site, bulldoze roads, dig slush pits, and drill around the clock. Prudent oilmen usually try to work with landowners in such matters involving surface usage, but they are not obliged to, which can be an unpleasant surprise to a surface owner who grabs a shotgun and tries to chase a work crew off his land. More to the point, the owner of only the surface of the land will not get a bonus or a royalty. The bonus and royalties will go to the owners or lessors of the minerals as long as they have not sold or otherwise parted with the rights.

On the heels of its dazzling sales to XTO and Chesapeake Energy, Four Sevens Oil Company faced the problem of where to turn next. By mid-2006, leasing and drilling had moved into town, and drillers were leasing town lots of a quarter acre, so-called "rooftop leasing." The older members of the Four Sevens group hesitated at this prospect, though Larry Brogdon had done his share of suburban leasing. Marty Searcy, Ross Matthews, and Dick Lowe's stepson Brad Cunningham, however, were ready to move ahead with a new venture, and since the thickest—and, hence, most potentially productive—part of the Barnett Shale included the area right under city streets, going into city leasing seemed the next step. At least, it would work if Four Sevens could manage to get right-of-way for a pipeline. Existing pipelines looped around, but not through, town. Luckily for the partners, the old Fort Worth and Western Railroad had a right-of-way from I-35 right through town and out west to Granbury, and they were able to purchase this splendid route for a pipeline. Searcy and Cunningham then launched a campaign to lease acreage on both sides of the right-of-way, part of a more general strategy to get an attractive corridor of urban land, drill some wells, and, repeating past history, sell out.[14]

Searcy and his partners knew that acquiring the number of town lot leases they needed would not be easy, but they were unprepared for how costly it would turn out to be. Virtually every home lot owner had to be approached for a mineral lease, and obtaining enough acreage to be worth drilling could mean negotiating hundreds of leases. When Searcy and Cunningham started leasing, they could pick up land for around $5,000 an acre, but the price rapidly escalated to $10,000 an acre or more! Homeowners learned to play rival landmen against one another. As Searcy described it,

> It didn't matter who you were trying to lease . . . they'd say, "Well, what's your number?" And then they'd go to company XYZ and ask them for their number, and . . . you were being ping ponged every day. What you end up doing is, come in and say, "Okay, we want to lease you, we will pay you a lease bonus of $X, but we want to sign the lease right now." 'Cause if you let them get away for twenty-four hours, there's another number that you had to jump over.

Once homeowners organized neighborhood associations, their gas committees dickered over bonuses and other issues. Before Searcy succeeded in leasing nine thousand acres, bonuses had risen to $15,000 an acre, and, as Searcy put it, "I was not sleeping very well."[15]

Searcy realized, moreover, that once drilling within the city got underway, if there was a blowout, fire, or other mishap, it would be catastrophic. The possible loss of lives or property would be the worst situation, but even if that didn't happen, it was reasonable to expect the city to respond to misadventure by shutting down all drilling within city limits. In April 2002, Four Sevens had to deal with a potentially dangerous situation when a plug blew out on a well it drilled in Haslet; thirty adults and children in a nearby daycare center were evacuated, and Four Sevens put up homeowners in a hotel over night until the repair was complete. No one was injured, but the incident demonstrated the hazards of neighborhood drilling. In April 2006, a wellhead malfunction on a drilling rig working for XTO in suburban Forest Hill resulted in a gas blast killing one worker and forcing the evacuation of five hundred nearby homes. Homes closest to the blast were only five hundred feet away from the rig. The Four Sevens group did put together some nine thousand acres in city neighborhoods, but in 2007, when Chesapeake made an offer to buy its leases and right-of-way for $250 million, Four Sevens did not hesitate to take the deal.[16]

If the Forest Hill blast gave some Fort Worth residents misgivings about welcoming drilling rigs into their neighborhoods, it did nothing to slow the headlong pace of leasing, not only on the part of homeowners and neighborhood associations but also by schools, churches, civic organizations, and local governments. By August 2003, Texas Woman's University and UNT had both leased part of their campuses and were drawing royalty checks; in one month UNT received "as much as $10,000" in royalties.[17] The Baptist Foundation of Texas and the American Cancer Society had both signed leases on donated land by the end of 2006; the Girl Scouts leased land under their summer camp. Bishop Kenneth B. Spears of the First St. John Baptist Church in southeast Fort Worth called a $21,000 signing bonus "manna from heaven."[18] In June 2006, the city of Irving leased land at its landfill, Twin Wells Golf Course, and numerous parks to Chesapeake; two years earlier the city of Fort Worth started leasing parkland.[19]

Fort Worth's move to lease city parkland, which was by no means universally applauded, prompted an unusual controversy over an eight-acre lease to Chesapeake Energy. The land was a park-like hiking and biking trail along the Trinity River, graced by large trees and not far from the TCU campus. Outraged that Chesapeake would drill on what would be called the "Trinity Trees Site," where she walked her Doberman named Barbie, Melissa Kohout organized a Labor Day rally against Chesapeake's permit in 2007. When the city did not revoke the permit, she hired a lawyer to represent herself and her dog, and ultimately sued the city on the grounds that it violated its own ordinance forbidding a drill site within six hundred feet of a city park. But to the surprise of many who hiked and biked along the tract, the city only owned a one-hundred-foot-long corridor along the river levee; the rest of the tract had been railroad land and zoned for heavy industry! Though the city had maintained the corridor and its trees, it had also issued permits along it for welding shops, temporary concrete batch plants, and power stations. In March 2008, State District Judge Bob McGrath dismissed Kohout's suit.[20] Chesapeake, however, did reduce the size of its drill site and cut down few trees.

The Trinity Trees controversy was but one example of what was more generally true: that Metroplex city governments were unprepared to handle the multitude of problems that surfaced when drilling rigs moved into town. Urban drilling was as new to local governments as it was to the homeowners who signed leases and waited for mailbox money. City codes did not cover the phenomenon; they did not even address the question of whether drilling in town was an issue to be handled by city councils or planning and zoning commissions. The initial response of some city governments, like the Denton City Council, was to prohibit drilling within city limits. In this instance, a coalition of companies put together by the Texas Alliance of Energy Producers pointed out to the Council that barring all drilling would be against Texas law; like it or not, the city had to allow mineral owners access to minerals.[21]

Similarly, Larry Brogdon encountered resistance when Four Sevens staked the first well within the city limits of Haslet. Never having been involved in drilling a well within city limits, Brogdon went to city hall to ask about rules and regulations. As Brogdon recalled, though Haslet had no ordinances regulating drilling, the city secretary said, "You can't drill a well in city limits." There followed meetings between the city council and Four Sevens's attorneys, resulting in the city reluctantly backing away from its prohibition. Brogdon reflected, "And that's when my political life started, I guess, because I got real involved in helping them develop an oil and gas ordinance."[22]

The main handicap shared by city governments was lack of an effective model for regulation. As Mike Moncrief, Fort Worth mayor during the boom, remarked, "There was no instruction manual . . . as to how to navigate the waters of an urban gas drilling program . . . one that could ensure safety to neighborhoods, make certain rigs didn't pop up in parks, and define where those gas lines were going to be run." For that matter, urban gas drilling was as much a novelty for oilmen as it was for city governments. Moncrief observed that industry members were used to drilling out in open country, "out in somebody's pasture." For farmers and ranchers, inconvenience and surface damage might be offset by a free cattle guard or a new stock tank; well sites might be at considerable distances from homes. Urban lessors would experience inconvenience, noise, and truck traffic essentially in their backyards. They had no use for cattle guards or stock tanks. And when city homeowners demanded intervention from city governments, industry members would confront new and unfamiliar rules. As Moncrief said of industry operators and their contractors,

> They're not used to being told when you can frack a well, what hours. They're not used to being told you will put noise dampers around your rigs. They're not used to being told that your water trucks are only going to be allowed to go this route. They're used to doing their thing, getting their well drilled, and moving on to the next location.[23]

In short, industry would have to do business in a radically different context.

Just as operators found themselves facing unfamiliar conditions, city governments confronted a host of challenging new issues having little in common with ordinary city business. Horizontal drilling would allow a number of wells to be drilled from one pad site and permit going under buildings and city streets, thus making it less invasive than conventional vertical wells. But exactly how far did well sites have to be from homes, schools, and other institutions to guarantee safety? Precisely how much noise would be beyond acceptable limits? How, if at all, should hours of operations be limited? Should frack fluid disposal wells be permitted within city limits? How would gas pipeline construction and location be regulated? And who would come up with answers to such pressing questions on such short notice? Adjusting new industry technology to an unconventional urban setting posed formidable problems.

Before rigs advanced into town, Fort Worth developed gas-drilling regulations that the City Council passed on December 11, 2001. Ordinance No. 14880 took on a range of issues including distance of well sites from various types of locations, noise, road damage, fencing in well sites, and camouflaging wells in neighborhoods. Drilling permits divided wells into three categories—rural, urban, and high impact—the last being a well within six hundred feet of a residence, religious institution, public building, school, hospital, or public park. Not only was six hundred feet double the setback distance required on other wells, but there were also more stringent restrictions on noise and delivery times for high-impact wells, and a permit for one required a City Council hearing.[24]

By 2006, as drilling rigs were moving into city neighborhoods, some residents questioned the adequacy of the 2001 ordinance as well as the wisdom of urban drilling. A resistance movement emerged, largely led single-handedly by environmental advocate Don Young, whose activism was prompted by the city's decision to allow drilling in the Tandy Hills Natural Area. A city park of roughly 160 acres, the Tandy Hills site was one of the few remaining areas of North Texas prairie filled with native plants and wildlife. As a youngster, Young played in the park, and he eventually came to live right across the street from it in the Meadowbrook neighborhood. In 2005, as he remembered, "I was standing at the park one day talking to a City Council person . . . and someone said, 'Hey, I heard they're going to drill over there' . . . That was my instant activist moment. I said, 'No way they're going to drill in my park here. This is my park, and I love it, and I'm going to take care of it.'"[25]

Young rounded up neighbors, first to form the Friends of Tandy Hills and then to organize Fort Worth Citizens Against Neighborhood Drilling Operations, FWCanDo. Energetic and imaginative, he began a valiant campaign to mobilize those unhappy with the prospect of urban drilling. He printed fliers and distributed them in his neighborhood, asking those who were like-minded to a meeting at his home. Much to his surprise, over two dozen people showed up to support him. He set up a website, collected information, developed a form for petitioning against drilling permits, and started showing up at both city council and neighborhood leasing meetings.[26]

Though the initial response from neighbors was encouraging, Young soon learned he had taken on a daunting crusade in which he would at first receive no help from national organizations like the Sierra Club or Nature Conservancy. He observed, "They knew somewhat about what was going on, but not much in those days. . . . So I was kind of on my own for a long time." When he showed up at leasing meetings to pass out anti-drilling fliers, he often faced harassment and threats from landmen organizers; he even received hate email. City officials did not welcome either his presence or input; when the city organized neighborhood meetings on changing the 2001 drilling ordinance, Young was told he could attend but not speak. Perhaps most discouraging, as lease-signing bonuses rose steadily, many residents seemed deaf to his anti-drilling campaign—those who did listen would respond with a "Can't fight Big Oil" attitude. Five years later, he would reflect, "I had no idea what I was getting into when I began this fight. If I had, I might have not done it."[27]

Like Young, however, enough Fort Worth residents were apprehensive about seeing a rig suddenly appear down the street and seeing home values fall as a consequence that Mayor Moncrief appointed a task force of gas industry members, real estate developers, and neighborhood association leaders in November 2005 to come up with a new drilling ordinance. As if to underscore the importance of the task force's work, the Forest Hill gas well blowout occurred as it was holding meetings, giving fresh anti-drilling ammunition to Don Young, whose FWCanDo pushed for requiring a three-thousand-foot distance between well sites and buildings. When the Task Force backed away from major changes in well location, Young declared its work nothing more than putting "lipstick on a poison pig."[28] Nonetheless, the forty-three-page ordinance the city council passed in June contained a great number of specific permit requirements, particularly for high-impact wells. These included limits on well site work hours (actual drilling excepted), heights of tanks, use of pits to store water, vehicle routes, and work-site noise. The ordinance even specified the color of the chain link fencing for high-impact or urban well sites—dark green or black—and the type of trees for site landscaping: no more than 25 percent evergreen. But with all this attention to detail, what could continue to concern neighborhood residents was the provision that so-called high impact wells—those near homes, schools, public buildings, and the like—could still have less than a six-hundred-foot separation from structures if such wells were approved in a public hearing or if property owners consented to them. The ordinance also did not address issues that would become controversial, like pipelines in neighborhoods or emissions from wells.[29] The ordinance certainly did not end controversy.

While Fort Worth city government struggled to come to terms with urban drilling, the gas boom proceeded at a feverish pace to the south in Johnson County. From 2004 onward, county residents saw an influx of hordes of landmen give way to an invasion of drilling and service company workers, as a drilling boom got underway in fields and pastures. Since wildcatters had generally ignored Johnson County in the past, the county had sustained a rural character. Now it would see a headlong rush to bring industry in the form of rigs, gas wells, and pipelines to the countryside, along with all the people necessary to put in and maintain them.

Perhaps most noticeable was the dramatic expansion of Cleburne's modest industrial base which, pre-boom, included a Walmart Distribution Center, Johns Manville, Rangaire, and Rubbermaid. Now a host of energy service companies came to town, including large firms like Schlumberger, Frac-Tech, Weatherford International, and Key Energy, as well as smaller players like Quality Logging and Chaulk Services. By the end of 2007, Cleburne Economic Development Foundation director Jerry Cash noted that within the past five years, 175 new businesses directly related to the jump in energy exploration had come to the Cleburne area. Incoming firms generally had to supply their own office space. Thus, Schlumberger bought a twenty-eight-acre site for offices and space for a hundred trucks; Quality Logging bought an acre of land, converted a four-bedroom mobile home into an office, and put five logging units to work. With a test facility at Alvarado, eight miles from Cleburne, oil field service company Halliburton solved both office space and workers' housing problems by putting up a field camp on a hundred-acre tract.[30]

Newcomers' need for places to live and eat immediately exceeded Cleburne's limited resources, and encouraged national motel and fast food chains to expand or locate in town. Booked 100 percent in September 2004, Cleburne's largest motel, the Comfort Inn, added twenty-six rooms, while other national chains, like Holiday Inn, Super 8, Home Studio Suites, LaQuinta, and Best Western, hurried to build facilities. With apartment occupancy running at 90 percent, developers also rushed to put up large new complexes of over a hundred units each. Even so, when Quality Logging came to town in 2004, the firm had to buy mobile homes to house the workers it had to bring in from Midland. Long lines of customers waited to take seats in local restaurants struggling to meet twenty-four-hour demand, prompting national fast food chains to open outlets along main thoroughfares. Nor was Cleburne the sole focus of explosive growth. Along Route 67, from Burleson on the northern county line south to Cleburne and beyond, open country gave way to a proliferation of fast-food drive-ins, trailer and RV lots, and pawnshops. Newcomers had money to spend—but sometimes they ran short before payday. As Diana Miller of the Johnson County Economic Development Commission summed things up at the end of 2007, "The growth is coming at us like a freight train."[31]

Explosive growth meant a proliferation of job opportunities for Johnson County residents. In 2007, economic analysts in the Perryman Group calculated that growth from Barnett Shale action had created 5,724 jobs in the county and generated over $300 million in annual personal income. The county's unemployment rate dipped as low as 3.5 percent in March 2008; basically, anyone who wanted to work could get a job. Especially in energy-related employment, employers competed with one another for workers.[32]

Oilmen competed for leases as well as labor, so just as in Fort Worth, lease bonuses jumped to new heights. David Arrington recalled that when he began leasing in Johnson County in 2003, most landowners were happy to accept bonuses of $125 an acre, "laughing under their breath all the way to the bank that some crazy guy from Midland had come to Johnson County where there's no oil and gas found and gave them $125 an acre." But particularly once Chesapeake began assembling acreage, bonuses rapidly escalated to stratospheric levels of as much as $10,000 an acre. As Arrington said, "It was crazy." Arrington thought Chesapeake's CEO Aubrey McClendon deliberately escalated bonus prices, running them up to the point where his competition was eliminated. Other observers, however, saw Arrington and McClendon as primary rivals in a neck-and-neck competition to lease. Jerry Cash remembered, "Between Arrington and Chesapeake it was kind of like a horse race, you know, whoever paid the most got to the finish line."[33] In any event, a tremendous number of Johnson County residents found themselves with an unexpected windfall in the form of bonuses and, later, royalty payments.

The city of Cleburne was a leading beneficiary of the wealth generated by the boom. It saw its property valuations rise from $1.3 billion in 2001 to $2.4 billion in 2010. Between 2000 and 2008, sales tax revenues rose from roughly $3 million to $8.8 million in 2008; in the three months ending in September 2008, the city also took in over $132,000 in hotel occupancy taxes. Since the city owned and leased a substantial amount of land, it also enjoyed a budgetary windfall from bonuses and royalties; royalty payments alone brought the city $375,000 in one month, February 2007. Even after gas prices dropped in 2009, the city still took in over $200,000 a month in royalty payments. Cleburne city leaders embarked on an ambitious program of capital expenditures—a new sports complex, a new building for the Chamber of Commerce and Cleburne Economic Development Foundation, renovation of the golf course, a history center, a recreation center, and a conference center. Similarly, revenue from leasing the Cleburne Industrial Park let the Economic Development Foundation purchase additional acreage for the park; gas royalties increased the Foundation's assets by $682,000 in 2008.[34]

Johnson County, however, owned far less land than Cleburne itself and thus did not benefit as handsomely from bonuses and royalties. Moreover, it had to cope with a problem less pressing within city limits. Many miles of roads had been damaged by a massive increase in traffic, particularly in the form of heavy oilfield equipment and tanker trucks. Just to drill and produce a well took hundreds of truckloads of drilling materials and frack fluids, not including trucks hauling used fluid to disposal wells, all going over country roads never intended for industrial traffic. Even state highways like Route 171 soon had awe-inspiring potholes. As a gesture of good will, some of the larger producers including Devon and Chesapeake voluntarily contributed to county repair work, and to encourage them, county commissioners mailed them estimates of the costs of fixing the roads the companies used most. The smaller contractors companies hired were less likely to share costs. In any event, contributions covered only part of costs. In 2007, for example, the county received over $800,000 from companies, but it spent over $1.4 million on roads and bridges. There was a silver lining, however: at the end of the 2006 fiscal year, the county ended up with $1.5 million more in its general fund than it had in 2005, prompting County Judge Roger Harmon to observe, "In my thirteen years on court, this is the best fund balance we've had in this county."[35] At that rate, perhaps it was worth putting up with some road damage.

Road damage was not the only downside of boom-time development. As contractors rushed to drill wells, not all heeded state drilling regulations or drilled responsibly. Texas regulation, for example, requires that pits used in drilling to store fluid or take brine must be lined so that fluid does not escape and pollute ground water. David Watts recalled that when he first got to Johnson County, "I got out at the well site, and they had a pit dug out there. There was no liner in it. And I'm going, 'Where's the EPA? Where are they at?' And half the wells I saw up there [the pits] didn't have a pit liner in them. . . . They were building locations so fast that they just . . . didn't bother."[36]

Nor were all operators considerate of landowners when it came to surface damages. One mineral owner leased a tract he intended for a residential development without making any provision in the lease for compensation for surface damage. The contractor hired by the lessee cleared ten to twelve acres of wooded land, destroying thousands of tall oak trees, removed part of a hill to level a pad site, and drilled a gas well in the middle of the thirty-acre tract the owner intended to develop. In a letter to the editor published December 16, 2007, Henry Rayburn, the tract's owner, told readers of the *Cleburne Times-Review*, "To us, they [the company leasing] were sloppy, careless and thoughtless, inconsiderate and wasteful. . . . Now our property has no value as a development." The company offered no compensation for damages.[37]

In short, in Johnson County, as elsewhere, drilling the Barnett bonanza emerged as a mixed blessing. It brought dramatic economic growth, thousands of new jobs, and unforeseen wealth to many people—but not without unfamiliar problems and stresses generated by industrial activity. By the end of 2007, Barnett leasing and drilling reached unprecedented levels, and the peak of the boom still lay ahead.

CHAPTER 5

From Boom to Bust

FROM A NATIONAL PERSPECTIVE, AT THE BEGINNING OF 2008 the economic outlook was anything but promising. The nationwide housing bubble had burst, plunging some of the country's largest banks into a sea of red ink and costing thousands of Americans their jobs. Yet, ironically, in January 2008 economic prospects for Fort Worth and the surrounding region could not have seemed brighter. As the regional gas boom gained even more momentum, it sparked an area-wide real estate boom. Investors poured money into the regional economy. The number of new jobs available skyrocketed. Area cities, schools, churches, and homeowners received hefty bonus and royalty checks in the mail. It was as if the Barnett Shale boom insulated Fort Worth from the impact of national downturn.[1]

Prompted by steadily rising oil and gas prices, drilling advanced into and under downtown Fort Worth. A headlong rush to lease a shrinking quantity of unleased urban acreage resulted in what amounted to bidding duels between energy companies, a fierce competition driving bonuses and royalties to levels unprecedented in industry history. And behind red-hot action was dramatic escalation of oil and gas prices, which would peak in July. Who could expect, under these circumstances, that downturn was even possible, let alone looming ahead?

With respect to growth in the Barnett Shale region, the statistics alone are astounding. During 2007, the total amount of dollars exchanged in the region jumped from over $16 billion to almost $23 billion, according to the economists in the Perryman Group. During the same year, the number of Barnett-generated jobs rose from almost sixty-one thousand to nearly eighty-four thousand. The energy industry, of course, offered opportunities to thousands of workers, but so did sectors supplying what workers needed and wanted. Thus, the biggest surge in new jobs came in retailing, with jobs in restaurants and bars in second place. No wonder observers like David Feehan, executive director of the International Downtown Association, remarked, "Fort Worth has set itself apart not just from every other city but from the world's top tier of cities. . . . You can expect Fort Worth's current growth to continue far into the future." Noting the multiplier effect energy industry operations had on regional growth, Barnett Shale Energy Education Council (BSEEC) director Ed Ireland agreed in February 2008: "Even better news is that we are just at the beginning stage of all this economic growth."[2]

In addition to pad sites, drilling rigs, and noisy gas compressor stations, the boom made itself evident in many ways. As companies set up regional headquarters in Fort Worth, they competed for office space downtown. One company alone, XTO, had offices in seven downtown buildings by May 2008; its biggest regional rival, Chesapeake, leased twenty-seven thousand square feet of the D. R. Horton building and bought the Pier 1 Imports building. Developers, counting on growing numbers of affluent downtown workers, hurried to supply upscale urban residential units. The Omni Fort Worth Hotel, for example, neared completion of a thirty-nine-story hotel topped by eighty-nine condo units whose starting price was $650,000; not far away, the landmark Texas and Pacific Railroad terminal was being transformed into twelve stories of luxury loft apartments. The market for luxury housing extended beyond downtown in a growing demand for homes valued at over $1 million; from one million-dollar home sale in 2002, sixteen such homes sold in 2006 and twenty-six in 2007. One realtor observed, "Fort Worth has broken through the million-dollar high-end market, and the Barnett Shale has helped that. . . . We're going to see the high-end market soar to new heights in 2008 in new developments and older neighborhoods."[3]

For that matter, according to Marty Travis, manager of the honky-tonk Billy Bob's Texas, a growing taste for the high end surfaced in patrons' liquor choices: "People are drinking a higher-quality product, whether they're going from a Seagram's to a Grey Goose vodka. Maybe they're celebrating their big paychecks working out on the Barnett field." Similarly, at the Two Bucks Beverage Center on Felix Street, salesperson Shelly Marquis observed, "Ever since the boom in gas drilling started, we've got these young blue-collar guys with name tags on their shirts coming in every payday and putting down $139 for a bottle of Remy Martin XO or $79 for 15-year old Bushnell's—or the $100 Patron . . . they could no more have afforded [that] a couple of years ago than I could now."[4]

Escalating urban drilling received heightened media notice, in part as energy companies worked to persuade hesitant mineral owners to lease home lots. By autumn 2007, the *Fort Worth Star-Telegram* gave the action daily coverage in a "blog" that tracked new drilling sites, neighborhood meetings, and in particular, what energy companies were offering as bonuses and royalties throughout the metropolitan area. They urged readers to send in questions, saying, "We'll do our best to respond" and adding, "We'd love to see a copy of your lease offer."[5]

In fact, neighborhood residents often had questions that went beyond simply finding out how much money they might get from energy companies. Not only had many never before seen a landman, but they had no idea of what drilling and production would entail. Would drilling under homes crack foundations or damage water and sewer lines? Would the drilling rig be likely to explode? How long would drilling go on? When they called City Hall, the Better Business Bureau, or the Chamber of Commerce, there was no one ready to answer their questions. For that reason, Fort Worth mayor Mike Moncrief prodded energy companies to establish some sort of informational body. Eight companies—Chesapeake, Chief, Dale Resources, Devon, Encana, EOG Resources, Four Sevens, and XTO—responded to establish the Barnett

Shale Energy Education Council (BSEEC), a nonprofit educational foundation, and hired former university professor Ed Ireland as executive director. Ireland would make public presentations, post articles on the BSEEC website, and offer input to city councils and planning and zoning boards that were trying to draft ordinances; he served on Moncrief's Natural Gas Task Force in 2008. In effect, Ireland would literally speak for the industry. As he put it, "I became available to speak at groups, any group that wanted me, whether it was four people or four hundred, I would go."[6]

Lest simply offering industry information fail to counter doubts about neighborhood drilling, the spring and summer of 2008 saw energy companies take up far more aggressive promotion of shale development, perhaps because determined environmental opposition, spearheaded by a persistent Don Young, seemed to be gaining momentum. Chesapeake Energy, for example, began running two-page advertisements in the *Star-Telegram* Sunday edition. Over a dioramic picture of the downtown Fort Worth skyline, flanked on the left by a drilling rig and on the right by a steer (acknowledging Fort Worth's historic title, "Cowtown"), ran a banner headline: "The Major Players in the Barnett Shale Helping Us All Win." Below, the text told readers that "Chesapeake Energy works with the best and brightest in the industry to bring the benefits of the Barnett Shale to you." And if the reader still failed to grasp the message, the text concluded, "Together WE ALL WIN."[7]

Not content with mere print, however, Chesapeake turned to television, and ran a half-hour infomercial dubbed "Citizens of the Shale" about gas production's economic benefits. That was accompanied by an ad campaign, both televised and in print, featuring Texan actor Tommy Lee Jones. Jones urged listeners to "Get behind the Barnett" and told them "The Barnett Shale is a national treasure that will benefit all Texans for generations." Chesapeake's public relations campaign reached new heights in July, when, as natural gas prices happened to peak, the company unveiled a plan to launch a web site solely about the Barnett action, Shale.TV. Although it never quite got off the ground, it signed popular KTVT newsman Tracy Rowlett to be anchor and managing director of the proposed online video channel that would include a talk show, public hearings, and panel discussions with industry experts. Though Chesapeake vice president Julie Wilson insisted that Shale.TV would be educational rather than simply advertising, the venture would nonetheless be produced by a subsidiary of advertising agency Ackerman McQueen. Mitchell Schnurman of the *Star-Telegram* called Shale.TV "a savvy way to soften up the public and build up good will." Less charitable observers derided the idea as "Shill.TV."[8]

One of the most novel features of the peak of the boom was the energy companies' inclination to lease and drill under anything and everything. Horizontal drilling, of course, was the technology permitting this development, since multiple lateral wellbores could be drilled from one well site and extended as much as two miles through the shale. Of all operators, Chesapeake was most venturesome in its urban projects. It drilled under the TCU football stadium, and, from seven pad sites, it drilled from the Fort Worth Zoo through to Southwestern Baptist Theological Seminary. Leasing nearly forty acres from the Bass family in May 2008, the first major downtown lease, Chesapeake prepared to drill under the Sundance Square entertainment district, the Sanger Lofts, and the D. R. Horton Tower. It acquired four pad sites outside downtown from which to drill.[9] Most striking, by the time it leased Sundance, it had already brought in production from under DFW Airport.

By using horizontal drilling and drilling multiple wells from one pad site, Trevor Rees-Jones's Chief Oil and Gas had shown it was feasible to drill under busy Alliance Airport, but drilling under the nation's third busiest airport was a project of far more daunting scope. Still, in the spring of 2006, DFW was an alluring target because its eighteen-thousand-acre lot was one of the few large sections over the Barnett Shale that remained unleased, and it was over one of the thickest, most promising parts of the shale. In March 2006, the DFW Airport Board sent out bidding packages to twelve of the larger regional operators, but once Chesapeake let it be known that they would offer top dollar to drill, potential rivals backed off, letting Chesapeake win the acreage with the offer of a $185 million bonus and a 25 percent royalty. The offer was typical of Chesapeake's over-the-top strategy, since the airport board expected only a 20 percent royalty. At the time, it was the highest bonus and royalty ever offered in the Barnett play.[10]

Having outbid potential competitors with its stunning offer, Chesapeake turned to the logistics of putting wellbores under runways. In order to avoid potential fracking hazards, seismic work was the first necessity, presenting the challenge of how vibroseis vehicles, which had to roll along thumping the surface to send sound waves through underlying rock, would function without shutting down the airport. Chesapeake hired Dawson Geophysical Company, which then had to negotiate a plan of operations with the airport bureaucracy: not only the airport board but also its several subordinate groups. Dawson's Ray Tobias reflected, "When we were awarded the program, I thought, well, at least the permit process will be easy, because it's all DFW property . . . it became very obvious very quickly that the process wasn't going to be easy . . . [because] you have to go to each individual group." The FAA, for example, wanted to be sure Dawson's work would not interfere with ground radar and required Dawson to undertake a week of testing performed between 1:00 and 3:00 a.m. With permits granted, Dawson began work, running between 11:00 p.m. and 6:00 a.m. The huge extent of the airport helped operations, since half the airport could shut down with enough accommodation remaining for night landings. The whole job took over three months. Tobias recalled, "We were glad when it was over."[11]

Chesapeake spudded in its first airport gas well on May 22, 2007, using a golden drill bit in honor of the occasion. Vertical wellbores went down to about 9,000 feet and then extended horizontally for 3,500 to 4,000 feet in the shale; five rigs, their height strictly regulated by the FAA, worked round the clock as the company proceeded to drill. October saw production under way, and by the following March the average daily production of gas was about fifty-six million cubic feet a day, with forty-seven wells completed of the three hundred planned. At the time, this meant a handsome return both for Chesapeake and the airport; the airport board intended to spend its bonus and royalty income on improvements to terminals.[12]

Like DFW Airport, schools, community colleges, park districts, and city governments faced the pleasant but demanding task of how to use windfalls from bonuses and royalties. A number of considerations could enter into their decision-making. One basic question was whether to use income from leasing and gas production as a substitute for tax dollars in meeting ordinary budgeted expenditure, thus giving taxpayers a break. Alternatively, they could direct their windfalls toward tangible, lasting improvements: one-time capital expenditures. Cleburne, for example, did a bit of both, spending on things like recreational facilities and a conference center, while also pouring some $1.5 million into the city's general fund to keep from raising tax rates. Another issue was whether to let the original use of land leased determine where income would go—for example, spending from bonuses and royalties earned on leased city parkland would be limited to improvements on parks.[13]

Public institutions came up with a wide variety of ways to spend windfalls from leasing and production, but the majority tried to put money into long-term improvement. Many had a lot of money to work with. Mansfield's parks corporation leased over four hundred acres for $342,000, which it directed toward new soccer goals, bleachers, lighting, and playground improvements. By March 2007 Tarrant County College district had banked $3.9 million from bonuses alone and planned to use it to set up a scholarship program. By that time, Tarrant Regional Water District had received $26 million in bonuses; it planned improvements on twenty-seven miles of levees in the district's floodway system. In 2008, the city of Fort Worth—which expected a windfall of close to $1 billion in leases and bonuses over the next twenty to thirty years—decided to split revenue, taking about half to set up an endowment in the form of a "permanent fund" with separate accounts within it for parks, airports, the water and sewer system, municipal golf courses, and the Fort Worth Nature Center.

With these separate accounts, as Mike Moncrief explained, bonuses from leasing a large park like Gateway Park would be shared with "little parks that . . . don't even have room to lease . . . but they deserve to benefit." The remainder would go into infrastructure, capital projects, and other long-term improvements.[14]

However public institutions decided to spend their bonuses and royalties, unexpected riches could make a big difference. The Everman Independent School District, for example, was one of the poorer districts in Tarrant County, with some 72 percent of its students qualifying for free or reduced-price lunch programs. By chance, however, the sixteen-square-mile district lay over one of the richest sections of the Barnett Shale. The district struggled to pay for its eight campuses out of its $34 million budget. But by March 2007, two XTO gas wells under district land brought the district $530,000 in royalty payments, leading superintendent Jeri Pfeifer to exclaim, "It's like winning the lottery!" Better yet, XTO planned to drill more wells under district land. With its windfall, the Everman school board decided to put income into an investment fund to build new schools and improve existing buildings, sports facilities, and playgrounds. For the low-income taxpayers of the blue-collar community, this was good news indeed.[15]

Cities and school districts, of course, had far more acres to lease than the average urban homeowner, but as neighborhood leasing gained momentum, homeowners looked forward to leasing largesse coming their way, which was a powerful incentive to sign on the dotted line. For the first time in Tarrant County, realtors saw home buyers and sellers haggle over including mineral rights in a sale. With lease bonuses quoted in figures per acre and royalties in percent of production, however, it was easy to overestimate the actual amount lessors were likely to receive. In May 2008, Gene Powell, publisher of the *Powell Barnett Shale Newsletter*, estimated that the owner of a 2.2-acre lot in Tarrant County might get close to $16,000 in royalties—but that was over a thirty-year period, assuming a royalty of 25 percent, a gas price of $6/Mcf, and a well with continuing production.[16] The bonus a lot owner received up front could very well be the biggest chunk of income the owner would see; as lease bonuses escalated to many thousands of dollars per acre, however, that could mean a handsome windfall in the short run.

Town lot leasing meant landmen needed to research literally thousands of land titles, which, by mid-2008, created a virtual title-searching frenzy at the Tarrant County Courthouse. The County Clerk's office was in the process of scanning records and putting them online, but work on that project was far from complete. A few years earlier the office might have seen thirty researchers in a day; now over 130 researchers a day were likely to appear. Landmen began to line up in front of the plot and deed records rooms before 7:30 a.m., rushing to use computers and microfilm readers when doors opened at 8:00 a.m. At times competition for machines was so intense that fights broke out between users. There was little County Clerk Suzanne Henderson and her staff could do; even if she bought more computers, she observed, the office lacked space for them. On the brighter side, records users paid a $5 records management and $5 archive fee to the office, bringing in $5.1 million in a year. Photocopies, at $1 per page, were expected to bring Tarrant County's general fund $2.3 million that fiscal year.[17]

Competition to see records was tame compared to what landmen would face to get signatures on leases, for, by mid-2008 in many parts of the Metroplex, the companies they leased for had to meet the demands of neighborhood associations. These associations might represent areas anywhere from several hundred to several thousand acres in extent. The Greater Meadowbrook Mineral Leasing Task Force, for example, included five neighborhoods covering an area of 3,252 acres. It was certainly easier for landmen to lease this alluringly large amount of acreage by dealing with the terms of the Task Force rather than with the 5,440 individual homeowners represented by it, but a coalition like this was capable of driving much harder bargains on bonuses and royalties and being far more demanding on other lease provisions. Representing twenty-one subdivisions and neighborhood associations, the 360 Northwest Coalition leased with Chesapeake only after leases included provisions for road usage, ground water protection, noise controls, pollution constraints, fencing and lighting of drill sites, and cleanup. Apparently better informed than many groups, the Coalition insisted that Chesapeake compensate lessors who had to pay their mortgage lenders for permission to lease, and it forbade the company from deducting transportation and gas processing costs from royalty payments. Later in the play, less sophisticated lessors would learn that deduction of costs could greatly reduce their royalty checks.[18]

Demands of neighborhood groups were not limited to leasing terms. Some persuaded companies to give their groups money in return for leasing agreements. XTO, for example, agreed to give each association within the Mansfield South property owners coalition $10,000; in southwest Fort Worth, the company agreed to donate $70,000 to neighborhood associations and improvement projects if it was able to lease at least 85 percent of its target area. The Mont Del Homeowners Association in Benbrook not only got XTO to give it $12,500 but made the company also agree to pay for a stone bench and install it near a local duck pond. Mayor Mike Moncrief openly encouraged energy companies to be generous to neighborhoods inconvenienced by drilling and production. Calling company representatives together, he suggested they place a priority on winning public goodwill:

> I said, "When you go in, instead of just your standard up front lease bonus or royalty or mineral rights approach . . . why don't you go in and say, 'Okay, here's what we want to offer you for your lease bonus, here's the royalty we're paying you. We want to do something else because we know this is going to be somewhat disruptive to the neighborhood. . . . We want to build a walking trail in your park, and we want to light it. And we might want to put some lights out in the parking lot.' Or, 'We want to start an endowment for a foundation that can offer scholarships.'"

Moncrief dubbed this route to public approval "the Fort Worth Way," and he won some of the larger companies over to his point of view.[19]

The rapidity with which bonus and royalty offers rose was the result of companies' having to meet property owners' expectations in order to compete successfully for acreage, and those expectations were based on information to be had daily in the *Fort Worth Star-Telegram*'s Barnett Shale blog or even minute by minute through cell phone. Property owners knew a lowball offer when they saw one. Thus, in January 2008, when a landman tried to interest residents of Arlington's Woodland West area in a bonus of $5,000 an acre, homeowners who had seen residents in an adjoining neighborhood get $15,500 felt insulted. As one resident described it, "Everyone was stunned. A couple were mad. Surely they didn't think we were stupid enough to accept $5,000 when everybody around us had already gotten fifteen-five." Those present voted unanimously to demand "a much better offer." And the embarrassed landman admitted, "There is somewhat of a discrepancy between the offers." [20]

Those who signed leases could face far more serious problems, however, than failing to get top dollar, especially if they did business with poorly financed fledgling companies. Developer Leonard Briscoe Sr.'s Glencrest Resources assembled some three thousand leases from residents of the Glencrest, Everman, Forest Hill, and Mansfield communities, promising bonus payments within thirty to forty-five days. Months passed, and lease signers received no checks: Briscoe leased without having raised the money necessary to cut checks. After lessors hired an attorney to launch a class action suit against Glencrest in November 2007, they finally began getting checks from the company in the spring of 2008. By that time two Dallas law firms, sensing Barnett litigation potential, put up a website, barnettshalelawsuit.com, to offer legal services to dissatisfied lessors. When boom gave way to bust, there would be growing demand for services firms like these offered.[21] Even before that, however, community residents sometimes felt that companies were not treating them fairly.

As keen as energy companies were to pick up acreage, they were not always willing to give way to neighborhood demands, and sometimes abandoned negotiation for business hardball that was close to coercion. Residents of the Coldwater Creek neighborhood in Southeast Arlington, for example, were offered only a $600 bonus per quarter acre, a 20 percent royalty, and a five-year lease term by Thunderbird Oil & Gas; company landmen threatened to stop all leasing in the area if they didn't reach their target number of leases. In October 2007, when members of the Brentwood-Oak Hills neighborhood association objected to Paloma Resources' plan to drill a well within six hundred feet of homes and tried passing out opposition pamphlets at a Paloma lease-signing event, Paloma representatives called police to disperse them; some months afterward Paloma's agents threatened that neighborhood residents' bonus payments would drop if leases weren't signed by March 1.[22]

Indeed, hardball could go beyond landmen's negotiations with neighborhoods: companies sometimes resisted tighter local drilling regulations. When the city of Southlake tried to toughen drilling regulations to require at least one thousand feet between wells and homes, schools, and offices, as well as to limit use of "frack ponds"—pits to store frack water, Chesapeake threatened to take its leasing elsewhere. As Chesapeake spokesperson Jerri Robbins put it, "Basically they're saying they don't want natural gas drilling with that ordinance. . . . Those cities that write friendlier ordinances will certainly reap the benefits from it, and we'll be there, active in their community."[23] Clearly, the attitude implied in Mayor Moncrief's "Fort Worth Way" did not extend to all situations.

In many respects, the Barnett Shale boom peaked in the summer of 2008. Throughout the spring, rising oil and natural gas prices helped heat action to a boiling point. On the New York Mercantile Exchange (NYMEX), natural gas futures rose from an average of $11.15/Mcf in early May to $12.82/Mcf in early June, peaking in July at $13.47/Mcf. At the same time commodities markets saw a wild surge in crude oil prices, which peaked at $147/barrel on July 11. Industry experts were at a loss to account for the record high prices, especially that of crude oil, since such usual price drivers as Middle Eastern politics or mounting demand were not part of the picture. One group of analysts blamed speculators; another analyst, economist Adam Sieminski, observed, "Stupidity can drive decisions." But however one explained prices, they were a powerful incentive to lease and drill as soon as possible. Perhaps with that in mind, Chesapeake and XTO took their bidding duel for leases to spectacular levels, leaving the next largest leasing competitor, Carrizo, far behind. From hovering around $25,000 an acre, bonuses rose to the $27,500—$30,000 threshold. The all-time high bonus offer went to residents of Marine Creek Estates, who leased their mineral rights for a bonus of $32,500 an acre in September. Ironically, by that time, gas prices had fallen below $8/Mcf. Caught up in the boom mentality of nowhere to go but up, many energy companies failed to notice deepening national economic crisis. More ominously, in the face of a contracting market, natural gas production had risen 9 percent in the first half of 2008.[24] The stage was set for bust.

Wall Street was first to decide the boom was over. By the end of the first week of October, stock prices of leading energy companies in the Barnett Shale tumbled, along with natural gas prices. Quicksilver Resources and Range Resources saw shares fall over 21 percent; XTO's shares dropped 10 percent and Chesapeake's dropped over 16 percent. Pipeline company Crosstex Energy saw shares which had traded as high as $35 fall to $16. Even Devon, the top Barnett gas producer, saw shares fall 11 percent. Industry observers could say that energy companies were simply caught by mounting national financial crisis, but no one could pretend short-run prospects looked good.[25]

Energy companies responded with a variety of cutbacks. Though they finished drilling wells in progress, they did not necessarily go forward with the expense of completing them to actually produce gas. Within a year, many companies cut back drastically the number of rigs they had working. Devon went from working thirty-nine Barnett rigs to eight, Chesapeake from forty to eighteen, and Quicksilver from fourteen to five. They pressured service companies to renegotiate costs. Devon and Chesapeake were among those who cut back production; as a Devon spokesperson put it in June 2009, "We see absolutely no reason to continue to drill at this time. . . . It's better to leave that gas in the ground and sell it next year, or in future years." Many companies, however, had to drill wells if they were to keep leases acquired at grossly inflated prices. Thus Chesapeake, with thousands of acres bought at top dollar, had the choice of drilling to keep leases or losing millions of dollars by letting undrilled leases expire. Worse yet, Chesapeake had borrowed heavily to fund its leasing splurge. By November it had started to raise money by selling sizeable interests in what it had elsewhere to BP and StatoilHydro, and there were rumors that the company might be taken over or go belly-up.[26]

At the very least, energy companies with bloated, overpriced lease inventories could back away from paying exorbitant sums for new leases. At the beginning of October 2008, the great leasing bubble burst. Bonus offers that were over $27,500 or more shrank to $5,000 an acre as top dollar. As John Baen, real estate professor at the University of North Texas, summed it up, "The party's over for now. . . . Commodity trends move in waves, but I don't think we'll ever see bonuses get that high again. It was insanity." Perhaps it was, but some potential lessors refused to see new reality. Having delayed signing leases with the expectation that better offers lay ahead, they turned up their noses at bonuses of $5,000 an acre. Thus, a Lake Worth miniature golf course owner of 5.5 acres, hoping for a $30,000 an acre bonus, turned down Chesapeake's $5,000 an acre offer, which fell far short of the $165,000 he expected. At the same time, the neighborhood spokesperson for the Southwest Fort Worth Alliance indicated that the group would take $5,000 an acre as "a starting point." "If it isn't," she said, "a lot of people won't want drilling in their neighborhood for so little in return." In fact, most of the holdouts faced signing for much less or not signing at all. Baen reflected, "They may be angry the rest of their lives, because this was the Lotto ticket they didn't scratch."[27]

Not only were the neighborhood associations that held out for higher bonuses out of luck, but some groups who thought they had a desirable offer saw companies back away from earlier deals. Vantage Energy, for example, told southwest Fort Worth homeowners that it would honor an agreed bonus of $27,500 an acre to those who had already signed leases, but it cancelled its offer to those who had not yet signed. XTO cancelled its offer of $24,000 an acre to residents of the

Timarron neighborhood. When Titan Operating backed away from a $25,000-per-acre offer to residents of the Bedford Colleyville Mineral Rights Coalition, fifty outraged Colleyville residents tried to pressure the Colleyville City Council to refuse to lease city land to Titan if Titan did not resume leasing from them on prior terms. Not only would the city forego a $7 million bonus and a $200,000 donation to city parks if it didn't lease to Titan, but City Attorney Matthew Boyle reminded the council it would face criminal charges if it passed up a measure benefitting the city in order to further financial interests of private individuals. The Council voted to lease to Titan. As Councilman Tony Licata explained to the angry audience, "I enjoy serving you, but I'm not going to jail for you."[28]

As leasing and drilling dropped off, job losses marked the next stage of the bust. Of course, the first group hit hard were landmen. Hundreds had been hired at day rates by firms like Dale Resources, which picked up leases for Chesapeake to meet unusual demand; as demand vanished, so did their jobs. The once-bustling deed records room at the Tarrant County Courthouse returned to a sedate pace of seeing twenty or so visitors a day. Some landmen relocated to other producing regions like the Haynesville, Marcellus, or Eagle Ford Shale areas; others took work negotiating pipeline right-of-way agreements through neighborhoods. Next to suffer job losses were blue-collar workers out in the field. In March 2009, with the price of natural gas hovering around $4/Mcf or less, the rig count in the Barnett Shale had fallen to fifty percent of what it was a year earlier. As roughnecks, mud loggers, and truck drivers left for other areas, job losses in the Barnett energy sector would reach forty thousand over the year. Nor were energy companies likely to bring workers back soon. S. P. "Chip" Johnson IV, CEO of Carrizo Oil & Gas, observed in April 2009, "We now have three rigs running, and frankly I wish we didn't have to run any rigs." Mounting unemployment also took its toll on the economy's real estate sector. By mid-December 2008, the residential foreclosure rate in the Dallas-Fort Worth area was up 17 percent from the previous year, while commercial foreclosures were up 32 percent. With the bloom off the Barnett boom, there was

nothing to shield the Metroplex from the mounting national economic crisis, and when it came to real estate, as one observer put it, "A freight train is coming down for landlords."[29]

Rising unemployment and the real estate downturn meant budget problems for city governments. Many went from debating how to spend unexpected bonus and royalty windfalls to struggling with budget shortfalls. Most cities did not experience the full impact of the bust until 2010, but by that time big bonuses were a distant memory, and with natural gas prices seldom above $4/Mcf, so were handsome royalty checks. Revenue from property taxes nosedived, and not only because of the real estate sector's difficulties: mineral owners' royalties were subject to local property taxes, and like cities, individual mineral owners received far smaller checks. When energy sector workers lost jobs or relocated to other regions, sales tax revenue fell. As Cleburne City Manager Rick Holden reflected on post-boom Cleburne, "Nobody is spending nights in the motels, they're not renting the conference center for their safety meetings any longer, they're not filling up the restaurants at lunch with workers. The corporations selling pipes and fittings have since moved and gone back to Midland." Cleburne tried to avoid raising local taxes by transferring some of its royalty income into its general fund, but after three years, it had to give up that strategy. In 2013, Holden said, "We're on our own and not relying on that money. . . . It was an artificial way to keep the tax rates stable."[30]

While cities struggled to make budgetary ends meet, temptation rose to tap gas windfall funds set aside for capital improvements. In some respects, cities' ability to make financial transfers out of such funds was limited. Funds from airport leasing had to go to airport improvement, and many cities had tied park leasing to park improvement. But energy-generated funds that were not so restricted offered a budgetary alternative to cutting services or raising tax rates. For example, the city of Arlington decided to transfer money from the gas-generated Arlington Tomorrow Fund to save its bus service running from Southwest Arlington to Fort Worth. In May 2010 Fort Worth City Manager

Dale Fesseler projected a $77 million shortfall in the coming year's budget. As an alternative to cutting back city programs and services, he favored backing away from the city's 2008 decision to direct property taxes on gas royalties into restricted funds, letting such revenue go to general use. School districts faced similar tough choices, since they, too, saw decreases in property tax revenue; the Arlington district faced pressure from the United Educators Association to tap gas money to avoid staffing cutbacks. Notwithstanding the best boom-time intentions, the bust changed financial priorities. Vickie Rodriguez, Euless finance director, observed, "The main goal has always been to use the gas drilling revenue to fund debt reduction, but that was before the economy kind of tanked."[31]

By April 2009 there were few optimists left in the Barnett. When the fourth annual Developing Unconventional Gas Conference took place in Fort Worth, a reporter covering the meeting decided that its key message was survival. Most saw the fortunes of natural gas as bleak for at least the rest of 2009; 2010 remained uncertain, but if the national economy stayed in the doldrums, so would the oversupplied market for natural gas. Having backed off leasing and drilling, Barnett Shale producers' next option, in the interest of weathering the downturn, was to sell off some of their Barnett assets. In May 2009, Denbury Resources arranged sale of 60 percent of its Barnett Shale gas assets to Talon Oil & Gas; Quicksilver Resources sold a 27.5 percent interest in its Alliance leases to the Italian petro giant Eni. In December Carrizo announced it would sell part of its Barnett assets to a subsidiary of Sumitomo Corporation. The most spectacular sale came at the end of December, when Exxon Mobil made a $40 billion offer for XTO; once the sale was finalized the following June, Exxon announced it would make the Barnett the focus of its unconventional natural gas operations.[32] In effect, when the largest American major company took a leading position in what had for so long been an independent's play, it marked the end of an era in regional development.

CHAPTER 6

Backlash

As the heady excitement of the Barnett boom subsided, an increasing number of persons had second thoughts about industry operations. They started to listen to longtime urban drilling opponents like Don Young, Gary Hogan, and Louis McBee, who emphasized the safety issues raised by drilling for gas in town. Neighborhood associations, no longer preoccupied with snaring attractive lease bonuses, shifted attention to problems that were an inevitable part of nearby drilling—noise, heavy truck traffic, nighttime lighting, pad sites adjoining homes and schools, and pipelines running through front yards. But it was area residents outside Fort Worth city limits who would come forward with far more damaging criticism of what the industry did. They would shift the focus of local opposition to drilling into a larger environmental context. Offering alarming accounts of injuries to their property and their health, they raised strident criticism of the Railroad Commission and the Texas Commission on Environmental Quality for failure to act on their complaints. Demands for action unmet, they pushed for responses on the part of their state legislators and federal authorities, in particular the EPA. With the election of Barack Obama, they found a newly aggressive EPA ready to listen to their complaints.

During the earlier stages of the boom, area residents seemed to have been focused on the prosperity generated by drilling and were relatively untroubled by what was going on around them. Sociologist Gene Theodori polled Wise and Johnson County residents on their reactions to boom time problems in 2006, following up with a similar poll in Tarrant County in 2009. The persons polled, like most Americans, had a negative perception of the petroleum industry in general but had a positive view of gas development, whose benefits they saw outweighing problems. Country dwellers, however, tended to be more negative about drilling than those in town; of those questioned, residents of largely rural Wise County were most negative, while Fort Worth residents living where a rig might be only a few blocks away were less upset by industry operations. True, there were those like Gary Hogan, who regretted leasing once they saw how aggravating the noise, dust, and traffic that came with drilling were, but they were generally outnumbered by those who saw little impact on their daily lives, like TCU English professor Dan Williams, who noticed the drilling taking place two blocks from his home at first but then found its impact "kind of negligible." TCU Provost Nowell Donovan found traffic on his commute was heavier than usual, but thought it more likely to be the result of area construction projects than neighborhood drilling.[1] Perhaps many city residents were so used to noise and traffic that nearby drilling made little difference in daily life.

Some months before the boom peaked, however, there were signs of growing public discontent with the industry's apparent lack of attention to neighborhood concerns and its heavy-handed tactics when it met with resistance. In September 2007, at a community meeting on Fort Worth's north side, City Councilman Sal Espino tried to tell residents how a local gas well would mean money to improve Rockwood Park. Residents fired back with complaints that drilling would mean noise, truck traffic, and safety hazards. Matthew Hudson, president of the Fort Worth League of Neighborhood Associations, told the *Star-Telegram* that gas companies were often uncooperative about meeting with neighborhood groups to negotiate differences. Thus when the Ridglea North Neighborhood Association tried to discuss plans for a well in its area, a company representative told them, "If we can't control the meeting, we don't want to be there." Signs reading "Just Say No to Urban Gas Drilling" appeared in more and more front yards. City code officers termed them "street spam" and yanked them out.[2]

One problem rapidly assuming critical proportions was that of disposing water used to frack wells by trucking it out of town. Fort Worth city government held back on permitting disposal wells—wells that could be used for injecting water down into the briny Ellenburger formation—within city limits until June 2006, when it granted a disposal well permit to be used for a well in the old Arc Park softball field, a permit eventually acquired by Chesapeake. The City Council, however, backed away from its decision and put a moratorium on additional disposal wells until April 2008. With plans for hundreds of new wells and the only disposal site in the city limits, Chesapeake intended to truck in wastewater from projects all over town, only to be met with opposition from both the city and Don Young's West Meadowbrook neighborhood, through which disposal trucks would travel. The city said it only wanted the site to be used for waste from wells in its immediate vicinity, not all over town. West Meadowbrook activists did not want the well used on any terms. Not only did they object to constant truck traffic, but they also voiced safety concerns about wastewater spillage and storage; wastewater stored in tanks before injection could contain traces of oil and gas, making tanks an explosion hazard. Such fears were not without reason: a wastewater storage tank in Johnson County, hit by lightning earlier in the year, exploded and burned for six hours.[3]

In December 2007, *Star-Telegram* journalist Bob Ray Sanders reflected on whether the wealth generated by drilling would offset damages already done or whether Fort Worth would end up with a "severe case of environmental heartburn." He observed:

> Despite attempts by some companies to camouflage the destructiveness of natural gas drilling, we've already seen a terrible

> scarring of the land, with large swaths being cleared for access to rigs and the laying of miles of pipeline.
>
> In several parts of town, I'm seeing the ugly industrial sites necessary to support this growing industry. . . . Large trucks are crowding and destroying streets that led to once-quiet neighborhoods. And many of us have just begun to learn that some of those massive mobile tanks are hauling waste water that we still haven't figured out how to handle. . . . I can't help but ask: while some people are getting richer during the great North Texas gas boom, is the land getting poorer?[4]

Of course, one way those with second thoughts might slow down urban drilling would be to persuade neighborhood residents not to sign leases. That was the strategy advanced by drilling opponent Liane Janovsky of the Ryan Place neighborhood, when she learned gas drillers might put down as many as eight wells adjacent to Ryan Place. As she put it, "I'm coming from the point of view of 'If we don't sign, they can't drill.'" She and her friends tried to persuade neighbors that signing leases would mean "having hundreds of water trucks coming into our neighborhoods [that] would destroy our streets, clog our traffic, and make our lives miserable. . . . And then there is the issue of potential danger if one of those wells explodes." But as journalist Peter Gorman observed, Janovsky and others like her were fighting "an uphill battle in a town that's gas crazy." Companies like XTO sweetened their offers to neighborhoods by promising light and sound restrictions, as well as escalating lease bonuses. Confronted with tempting bonuses, residents could justify signing a lease by reasoning that if their neighbors signed and they didn't, drilling would happen anyway, and they'd see no "mailbox money."[5]

Rather than trying to head off leasing, a more feasible strategy to counter apprehension about neighborhood drilling and insensitive company project planning lay in pushing for a greater measure of city drilling regulation, particularly with respect to neighborhood impacts. City Council member Wendy Davis suggested in September 2007 that the 2006 ordinance needed review to strike a better balance between permitting drilling and protecting neighborhoods. Davis and City Planning Director Fernando Costa advocated reorganizing the task force responsible for the 2006 measure, particularly to tighten up restrictions on "high-impact wells" that would be located within six hundred feet of homes, schools, churches, and other public buildings. Davis and Costa's proposal initially met a chilly response from Mayor Mike Moncrief, but sustained opposition would place him in an awkward position, for as *Star-Telegram* reporter Mike Lee noted, the mayor drew a handsome income from his oil and gas interests. In the end, he gave a lukewarm endorsement to review, explaining, "I'm certainly not averse to reviewing what we have, how well it works or where it might have failed." However, he did not reappoint a task force until February 2008.[6]

The eighteen-member Natural Gas Drilling Task Force consisted of a representative from each city district, four industry representatives from Devon, Chesapeake, Quicksilver Resources, and XTO, and members representing developers and business interests. Although it included Gary Hogan, Jim Bradbury, and Susan De Los Santos—members who questioned the benefits of urban drilling—it was outnumbered by those likely to be industry friendly, including city staffers who would assist deliberations. Its agenda included revisiting old issues such as noise from drilling and distance from "protected properties" like homes, schools, and hospitals, as well as new concerns like pipeline routing and compressor station operations, but the task force was clearly designed to facilitate rather than hinder drilling more wells. Drilling opponents could therefore be skeptical that the regulation it might advance would be any more to their liking than the 2006 ordinance. Pressuring the task force to consider their concerns, particularly with respect to the safety of gas operations in neighborhoods, would require uniting industry doubters from a variety of quarters. To do so, former City Council member Cathy Hirt, who unsuccessfully opposed Mike Moncrief in the mayoral race of 2003, came forward to organize a new group, the Coalition for a Reformed Drilling Ordinance (CREDO) in late July.[7]

CREDO offered the opportunity to bring together a number of anti-industry activists. At its organizational meeting on July 24, representatives from the League of Women Voters, Citizens for Responsible Government, and the Sierra Club attended, as did Don Young. State Representative Lon Burnam offered his support, commenting, "Municipal government could do more but has failed to do so. We need to stop what we're doing and assess the situation." In fact, that was CREDO's first move, one guaranteed to get attention: it demanded a moratorium on issuing any new drilling permits. To drive its point home, its logo had "Whoa" printed above "Moratorium Now!" Hirt and others specifically singled out for criticism Chesapeake's massive pro-drilling campaign suggesting it would backfire by bringing the uncommitted to CREDO's point of view. As she put it, "How many people said anything until they saw Tommy Lee Jones telling them they needed to be citizens of the shale and that meant supporting the Barnett Shale unconditionally? You can say it the Shakespeare way: 'Methinks the fool doth protest too much.' Or you can say it the Texas way: 'Don't pee on my leg and call it rain.'" To rally support from those who shared their dissent, CREDO held a rally outside Will Rogers Coliseum on August 7. A crowd of over two hundred turned out, shouting, "Hell no! Just say whoa!"[8]

By the time CREDO held its rally, pipeline construction through city neighborhoods was well underway, and heavy-handed property condemnations along Carter Avenue, part of the East Meadowbrook neighborhood, sparked residents' outrage. Here was a problem Barnett area landowners outside city limits had already confronted. They had seen trees bulldozed down to cut construction corridors twenty to fifty feet wide across their land, corridors that hence forward would be restricted in terms of the landowners' own use. Pipeline companies paid landowners for such easements, and some owners were willing to sell. But owners who did not wish to part with their land came up against the unpleasant reality of Texas law: gas pipelines were considered public utilities, and pipeline companies could condemn property they needed by virtue of eminent domain. The Carter Avenue neighborhood was one of modest homes housing a mixed population of blue-collar

workers, retirees, and immigrants, many of whom were not in an economic position to hire the kind of legal help one would need to fight a gas company, which in the instance of Carter Avenue, was a Chesapeake subsidiary, Texas Midstream Gas Services. Texas Midstream initially planned to put a twenty-four-inch pipeline with a thirty-foot easement through the front yards of forty-four homes, thus virtually taking most homes' entire front yards. When homeowners resisted, Chesapeake began filing condemnation orders in May 2008.[9]

Notwithstanding Chesapeake's expensive public relations campaign, Carter Avenue resistance grabbed the headlines, in part because an unlikely poster child for opposition appeared in the form of seventy-two-year-old widow and retired artist, Jerry Horton. Having reduced its demand for easement from thirty to twenty feet, Chesapeake offered Horton $12,987 for its corridor across her front yard only to see her turn the offer down: she was not willing to let the company cut down the century-old oak trees fronting her home. Threatened with condemnation, Horton decided to fight city hall by appearing at a city council meeting to protest the line. She enlivened the question time concluding the meeting by getting into a shouting match with Councilman Jungus Jordan and the mayor, who said they were protecting neighborhoods. "You have not supported us," she charged; "They're gonna take my front yard and they're going to take your front yard, too, and they're going to take this whole city." Her tirade did not prevent her receiving condemnation papers three days later, but it gained her instant city celebrity, as well as national news coverage. Three weeks later, Horton decided not to extend the battlefield to court. In return for Chesapeake agreeing to put the line at least twenty feet deep, replace trees that died, and pay $15,500, she signed off, remarking "I am heartbroken," to which one *Star-Telegram* reader posted the comment, "This makes me wonder what kind of pressure and threats she was under by Texas Midstream and Chesapeake. I bet they made Ms. Horton an offer she 'BETTER NOT' refuse."[10]

Not only did Chesapeake confront an avalanche of negative publicity over its Carter Avenue project, but as controversy continued into the fall, city council members and some state legislators came out on the side of residents. The company couldn't expect wholehearted cooperation from City Hall. Reassessing its position, it decided to approach the Texas Department of Transportation (TxDOT) about laying its line along the I-30 right-of-way. TxDOT initially balked, citing safety concerns that had not troubled Chesapeake on Carter Avenue. At the end of November, TxDOT decided on negotiation with the company. Reluctant to admit complete retreat, Chesapeake continued to press condemnation proceedings against the lone Carter Avenue resident who was still a holdout: Steve Doeung. In February 2009, Wendy Davis, newly elected to the State Senate, introduced a bill requiring TxDOT to allow gas gathering lines in freeway rights-of-way; Governor Rick Perry's veto of it left Carter Avenue in pipeline limbo until the following year, when in March 2010, TxDOT, under pressure from Davis and Lon Burnam, announced it could approve Chesapeake's line once the company gave it detailed plans for the line. One by-product of the controversy was the founding of yet another citizens' group aimed at controlling drilling, the North Central Texas Communities Alliance, organized by Esther McElfish and Louis McBee. It took its motto from Calvin Tillman, mayor of the Denton County town of Dish (officially called DISH): "Together we bargain, divided we beg." After Carter Avenue, giving gas companies whatever they wanted was no longer in style.[11]

While Carter Avenue took media attention, during the summer and fall of 2008, the Natural Gas Drilling Task Force continued its toil on a new drilling ordinance, but not without a host of difficulties. One problem lay in what aspects of drilling and pipelines the city might regulate, given that the Railroad Commission was the regulatory agency for the oil and gas industry. Then there was the issue of how near homes and public buildings pad sites and wells might be, a matter industry representatives wanted to see left as open-ended as possible. This question made obvious the basic division in the task force, between those who wanted minimal regulation and those who wished for much more regulation than the 2006 ordinance provided.

Division would result in the latter group of Hogan, Bradbury, and De Los Santos delivering a minority report to the city staffers who would write up the new ordinance based on task force recommendations. City staffers ended up surprising both parties in the task force by adopting many minority report recommendations for the new ordinance. De Los Santos observed, "We were pleased that so much of the minority report was mirrored by what the city staff adopted . . . the gas companies were not happy." Jim Bradbury put it more succinctly: "Chesapeake was furious."[12]

With more at stake in city drilling than any other company, it was not surprising that Chesapeake found even the relatively modest limitations of the proposed ordinance unacceptable. What it did in response, however, resulted in another gaffe in public relations. The city scheduled a town hall discussion of the ordinance on November 20. Chesapeake packed both the meeting hall and parking lot with over 800 of its supporters—oil field workers, contractors, and others. Interested citizens were thus unable to attend and take part, and the meeting adjourned after only forty minutes.[13] In effect, Chesapeake enhanced the growing perception of an over-powerful gas industry determined to have its way in the community, the kind of "them against us" division that activists like Don Young and Gary Hogan had been talking about for the past three years.

While the ordinance the City Council finally voted for in December involved much more regulation than Chesapeake wanted, for those who questioned urban drilling, the measure did not go far enough—and, of course, by allowing drilling to continue, it was not acceptable to Fort Worth CanDo or to CREDO. Among the new restrictions in the ordinance was tighter noise control: before drilling, operators would have to take a seventy-two-hour measurement of existing noise levels, to be used as a baseline for future noise measurement, and file a noise abatement plan. Operators would have to use a closed-loop mud system on wells, one in which liquids and solids would separate in tanks and drilling waste would not be stored in open pits. That continual bone of contention, the distance of a well from homes and public buildings, was kept at six hundred feet, but which buildings would be in the "public" category remained controversial: jails and hotels were added to the list, but stores, restaurants, and movie theaters were not included. There was entirely new regulation of compressor stations, whose equipment had to be enclosed in buildings to cut down on noise and emissions; large compressor operations had to be sited in industrially or agriculturally zoned areas. Also new was the requirement that pipeline operators file construction maps and plans with the city, and there would be a Gas Drilling Review Committee to evaluate drilling permits. City regulation in both these respects, however, left hazy where city authority ended and Railroad Commission regulation began. The ordinance did not deal with the burdensome problem of frack wastewater disposal; tank trucks full of wastewater would continue to rumble through city neighborhoods. The ordinance, then, was far from ideal, but it did offer more protection from industrial disturbance to city residents. While other cities like Midland would look to it as a model,[14] Fort Worth would amend it the following year.

Outside Fort Worth, other communities worked on their own city drilling ordinances. Like Fort Worth, they struggled to strike a balance between gas companies and residents on the question of the proper distance between well sites and homes and public buildings. Arlington, Euless, Hurst, and North Richland Hills decided that six hundred feet was an acceptable minimum distance, though all but Euless provided for the distance to be lowered to three hundred feet with consent of property owners. By contrast, Colleyville, Grapevine, and Flower Mound insisted on a distance of one thousand feet; Flower Mound had passed its regulation as early as 2003 and would allow a variance only to a minimum of five hundred feet from homes, schools, and other public structures, which tended to discourage drilling in town. In January 2008, Bedford decided that it would permit drilling within six hundred feet of homes, parks, and many other buildings, but insisted on a thousand-foot buffer between wells and schools. The same year, Southlake decided on a thousand-foot limit between wells and homes, schools, and offices, but it also placed stringent limitations on water storage

tanks and frack ponds, prompting Chesapeake to respond that it would drop plans to drill in Southlake. Southlake also opted to charge more than most for a drilling permit: $15,000 a well as opposed to the more usual $3,000–$4,000.[15] Because state law guaranteed mineral owners their right of access to minerals, however, any of these municipal ordinances might be challenged in court by frustrated drillers. It was hard to say exactly how much protection they would offer.

While city residents could push for ordinances regulating what gas producers could do in their neighborhoods, those living outside city limits had few safeguards against industry operations they found objectionable. Many persons living over the Barnett Shale were country folk—or trying to be. A half-hour drive from downtown Fort Worth to Denton, Wise, Parker, or Johnson Counties took one to a rural world of hilly pastures punctuated with grazing cattle, groves of trees, and banks of wildflowers. This was the sort of environment where one could hope to escape from the noise and stress of urban life, where one could keep horses and other animals, where one could breathe clean country air. In this apparently idyllic environment, few things would be less welcome than the appearance of industry in the form of drilling rigs and compressor stations. Industry not only put a period to rural dreaming but also represented a threat to the whole environment. The industry presence could be seen as causing air and water pollution that threatened to kill animals, destroy vegetation, and sicken people. It was no accident that some of the most outspoken and vehement of industry critics came from the country. The first environmental issue they raised was air pollution.

By focusing on air pollution, industry critics not only found an issue that was certainly a timely concern given urban drilling, but they could also target a subject that had recently gotten extensive media attention. Late in January 2009, the Environmental Defense Fund released a report on air pollution in the North Texas area done by Southern Methodist University (SMU) engineering professor Al Armendariz. He had looked at nine counties in the Dallas-Fort Worth metropolitan area—Tarrant, Denton, Parker, Johnson, Ellis, Collin, Dallas, Rockwall, and Kaufman. Additionally, he looked at emissions from all oil and gas operations in counties with the majority of Barnett Shale gas wells—Tarrant, Denton, Wise, Parker, Johnson, and Hood Counties. Concentrating particularly on emissions from storage tanks and gas compressors, he calculated how much each tank and compressor would emit in fumes and multiplied that amount by the number of tanks and compressors in his region. He found that, in his whole study area, gas industry operations were responsible for adding an average of 191 tons of pollutants to the air per day. In the summer, emission levels rose, reaching 165 tons per day in a five-county metropolitan area with significant production. In the same area, automobiles and other vehicles were only responsible for an average of 121 tons of pollutants per day. He concluded that gas industry operations "likely [have] greater emissions than motor vehicles in these counties." The emissions included toxic chemicals like benzene and formaldehyde, chiefly from condensate tanks and engine exhausts, but greenhouse gasses like carbon dioxide and methane were also present.[16]

Now, no one could pretend that the Metroplex area had clean air. In fact, in 1990, a decade before Barnett Shale drilling was important, the EPA classified nine counties surrounding Dallas and Fort Worth as failing to meet national air quality standards. In 1999, the EPA classified air pollution in the area as serious. Armendariz himself had first studied air pollution from regional cement plants for a North Texas environmental group, Downwinders at Risk. Moreover, normal oil and gas industry operations did include some sources of air pollution. Exhaust from engines on drilling rigs and compressor stations was usual. Tanks storing condensate commonly released volatile organic compounds (VOCs) in small amounts. During gas well completion, some gas was ordinarily vented to the air. In fact, studies of gas field emissions done by the TCEQ in 2006 and 2007 offered some of the data Armendariz used. The shock value in Armendariz's report came from its emphasis on how much air pollution came from gas industry operations and its identification of toxic chemicals like benzene, a known carcinogen, in the air.[17] There lay cause for alarm. All industry critics needed was a graphic example of toxic pollution at work. One surfaced in May 2009.

Deborah Rogers had turned a forty-five acre farm five miles west of Fort Worth in Westworth Village into an organic dairy farm. On the farm, which had been in her family for three generations, she raised goats, chickens, turkeys, and some peacocks; she made prizewinning cheese from the goats' milk. In May 2009, Chesapeake bulldozed down trees adjoining her property line and began drilling gas wells. Rig engines ran day and night, and gas Chesapeake vented as it drilled made the air smell like gasoline. After several days, baby goats and chicks began to die, and Rogers herself had severe headaches and nosebleeds. Suspecting air pollution as the cause of problems, Rogers contacted the Texas Commission on Environmental Quality, which referred her to the Railroad Commission, which referred her back to the TCEQ. Frustrated by agency buck-passing, Rogers hired Alisa Rich's Wolf Eagle Environmental Engineers and Consultants to test the air on her farm. Wolf Eagle ran tests on several days, including one when Chesapeake flared gas. They found "the presence of recognized and suspected human carcinogens in fugitive air emissions present on [the] client's property commonly known to emanate from industrial processes directly related to the natural gas industrial processes of exploration, drilling, flaring, and compression." Among the toxic chemicals they identified as present in excess of the TCEQ effective screening levels (ESLs) were chloroform and carbon disulfide, and their report included a paragraph on each chemical's damage to human health. They also identified small amounts of benzene. Without considering alternative sources for the toxic chemicals it identified, Wolf Eagle concluded, "The presence of these compounds in a residential area indicates a strong correlation to gas flaring from gas exploration in process during the week of air testing."[18] Surely Rogers now had, in effect, a smoking gun for the TCEQ.

The TCEQ, however, did not take Wolf Eagle's testing as cause for alarm. During the next eighteen months they did their own air testing, but they did not find quantities of toxic substances high enough to pose a long-term risk to health. Rogers did some air testing herself and found carbon disulfide, prompting her to write Al Armendariz; he answered that the levels of carbon disulfide in her air were three hundred times what one might normally expect to find in an urban area, let alone a country farm. Chesapeake responded to her complaints by doing its own air testing. Not only did it reject Wolf Eagle's findings but insisted its operations could not be responsible for whatever might be in Rogers's air, for when Wolf Eagle ran tests, the wind was not blowing from its wells to the farm. Its environmental engineer, Grover Campbell, insisted, moreover, that gas flaring burned off toxic compounds and did not produce carbon disulfide.[19] Depending on whom one chose to believe, the air on Rogers's farm was either lethal or harmless.

If the TCEQ was inclined to dismiss Wolf Eagle's findings, the same could not be said of the Fort Worth City Council. Were Wolf Eagle correct, gas wells in city neighborhoods might be spewing carcinogens. The city decided to hire an evaluation of the Wolf Eagle report by a Carrollton firm, Industrial Hygiene and Safety Technology, Inc. (IHST). Its evaluation, presented late in August 2009, offered scathing criticism of Wolf Eagle's methods, which it called "rudimentary in scope and design," and results that IHST called "inconclusive at best." Wolf Eagle had not looked beyond gas well operations for sources of contaminants. Its presentation of the hazards of chemicals it identified, moreover, was "generally exaggerated and speculative, not representative of the hazards posed by the actual concentrations of compounds detected," and it misapplied TCEQ effective screening levels. Overall, the Wolf Eagle report, as well as letters from Rogers and Armendariz, did not offer sufficient evidence to show an "adverse impact" from gas drilling. Presumably, Fort Worth residents could now breathe more easily—though IHST did add that gas well drilling and production might have significant environmental impacts deserving investigation, especially "as the density of such operations increases."[20] In other words, the dangers that Wolf Eagle failed to prove might still exist.

While the Fort Worth City Council and the TCEQ might take temporary comfort in Wolf Eagle's report being, apparently, bad science, the Barnett Shale Energy Education Council developed a defense against charges that gas drilling caused air pollution. Starting with Armendariz's report, it faulted him for assuming that natural gas produced anywhere in the Barnett Shale was the same in chemical composition, yielding the same quantity of volatile organic compounds if vented. True, gas installations like those in parts of Wise, Denton, and Parker Counties handled a significant amount of "wet gas" containing a high proportion of liquids to be stored and volatile organic compounds like benzene, but gas produced within Fort Worth city limits—or on Rogers's farm—was dry gas with no liquids, virtually no benzene or other volatile organic compounds; therefore, no problem.[21]

Soon after the disturbing news about the air on Deborah Rogers's farm, upsetting reports of air pollution came from another rural quarter—this one from the Denton County hamlet of Dish. Little more than a wide spot in the road, Dish is a community of some two hundred people, west of Denton and between the slightly larger towns of Ponder and Justin. Beginning in 2005, Dish became the site of extraordinary development by gas companies. Originally named Clark, the community changed its name when a satellite television company offered it ten years' free television service for the change. Though tiny, Dish was a town with a mayor. In 2005 pipeline companies began getting easements from local landowners, and late in the year, Atmos Energy built the first compressor station just outside city limits. Soon after, Atmos built a second compressor station; Energy Transfer Partners (ETP) built two; Crosstex Energy (later EnLink Midstream) built one; and Enbridge Energy Partners LP and Texas Midstream Gas Services LLC each built three.[22] In an area with acres and acres of open country, why the gas companies decided to put their facilities right next to a small town is hard to understand. In little less than a year, Dish became a gas transportation hub, sending millions of cubic feet of gas out of the region. Dish residents soon began to resent their new industrial neighbors.

On the night of October 16, 2006, an Atmos compressor vented natural gas into the air for forty-five minutes; a buildup of pressure inside a pipeline tripped a relief valve to release the gas. Fortunately, the gas did not ignite—the release took place not long before midnight, and no one had to evacuate. But this was the second such incident that year, and since no employees were locally resident, it took company staff forty-five minutes to respond, during which time a Justin firefighter figured out how to shut down the compressor. Though the incident was minor enough not to be reportable by state or federal standards, it prompted Dish residents to ask gas company representatives to answer questions at a town meeting a month later. Representatives of only two of the companies, Atmos and ETP, bothered to appear. Atmos tried to reassure residents that it had changed valve settings and that there should be no future problems; the company was also working on better communications with police and fire departments. Residents, however, also raised concerns about noise and fumes coming from the installations. The company men promised that they would work on noise mitigation and use sound absorption devices, but they had far less to say about fumes beyond promising to check for small leaks. Dish mayor Bill Merritt said he would take them at their word. Calvin Tillman, however, threatened to complain to state authorities if the companies did not resolve problems. Once mayor, Tillman made good on his threat.[23]

In 2009, despite reassuring words from gas company representatives, problems with noise and foul odors still plagued Dish residents. One property owner, Jim Caplinger, whose home was only about 800 feet from the compressors, observed, "You don't want to go to the front porch for sure because it's so loud, and if you go to the back porch, well, the sound bounces off the trees out back and roars at you like a jet engine getting ready for take-off—but one that never takes off, just continually roars."[24] Not only did residents find the odors overwhelming, but the fumes also stung their eyes and irritated their noses and throats. Fumes seemed to have a visual impact as well, since trees in the town started dying. Dish resident Bill Cisco noted, "Those trees breathe air just like we do, so when they start dying, you've got to pay attention." Mayor Tillman complained to both the RRC and the TCEQ, but like Rogers, he found state agencies unhelpful: as he put it, the agencies "basically asked the companies to investigate themselves . . . and they came up clean every time." In May, the companies hired a firm to look for gas leaks, but the firm found none. Then Tillman said he received an anonymous phone call telling him to hire a firm to look for toxins in the air. The city decided to spend $10,000 to have Wolf Eagle analyze the air.[25] Thus, one might say, the torch was passed from Rogers's farm.

Wolf Eagle's report, presented on September 15, 2009, was well positioned to cause alarm. In fact, the first paragraph of the report informed the reader that, "Development of the gas industry in a residential area can cause degradation of Quality of Life (QOL), general annoyance to human populations, as well as noise, air, water, soil contamination, and vibrational disturbances of foundations of buildings."[26] Much worse news followed. On August 17, Wolf Eagle had monitored the air for twenty-four hours (using canisters with a twenty-four-hour flow) in seven locations on homeowners' land. It found many toxic chemicals in the air, including benzene, xylene, carbon disulfide, naphthalene, and dimethyl disulfide, at levels exceeding what the TCEQ set as ESLs—the level of a chemical suggesting the need for a review in greater depth, which might determine short- and long-term damage to health from exposure. As in the report for Rogers, Wolf Eagle offered a paragraph on each chemical, describing its toxic potential. Among the health problems the chemicals could cause were anemia, leukemia, renal cell carcinoma, kidney injury, nerve damage, cataracts, nausea, vomiting, diarrhea, jaundice, headaches, and irritation of the eyes, nose, and skin. And the report concluded the air in Dish contained "high concentrations of carcinogenic and neurotoxin compounds."[27] Using the same language as the Rogers report, it warned, "Many of these compounds verified by laboratory analysis were metabolites of known human carcinogens and exceeded both Short-term and Long-term Effective [sic] Screening Levels (ESL) according to TCEQ regulations."[28] At first reading, the warning sounded like breathing Dish air could kill you—or at least threaten your life.

Perhaps because Wolf Eagle anticipated a hostile industry response to its report or perhaps because Alisa Rich had only a master's degree in public health from UNT (and was working on a doctorate at the University of Texas at Arlington [UTA]), Rich summoned up reinforcement for her air analysis from Dr. Wilma Subra. Subra, a chemist and a member of the EPA's National Advisory Council for Environmental Policy and Technology, was active in the Earthworks' Oil & Gas Accountability Project (OGAP). Not only did Subra give a positive evaluation of Wolf Eagle's findings, but she would launch a health survey of Dish residents for Earthworks' OGAP in October and November 2009.[29] Presumably, the report would determine if the toxic chemicals had actually made anyone sick.

Subra's report, released in December, was not likely to allay anyone's fears about living in Dish. She asked the thirty-one persons who participated in her survey if they noticed frequent "odor events," or times when industrial fumes or other odors were annoyingly strong, what their general health was like, and whether these "odor events" seemed to cause health problems. The health problems that surfaced ranged from bronchitis, allergies, and eye irritation to brain disorders, nervous system impacts, and pre-cancerous lesions. But, emphasizing the frightening potential of Dish's air contamination in bold type, the report stated, "61% [of] Health effects reported by the Dish community were associated with toxics measured in excess of TCEQ screening levels."[30] This statistic certainly suggested that the airborne chemicals Wolf Eagle found were making people sick. One could, however, take issue with that assumption. Many of the ailments Subra identified could have been caused by things unrelated to toxic chemicals—depression, fatigue, and allergies, for example.[31] It may be worth noting that eighteen of thirty-one were over age forty; eight of her subjects were also smokers who had, on average, smoked for twenty-three years, and all of these said they had respiratory problems. A fifty-five-year-old woman who had smoked for over thirty years reported forty-six medical symptoms, though she said she was healthy.[32] On close examination, Subra's report did not explain why some Dish residents got sick—but it certainly implied that toxic chemicals in the air were at fault. Just looking at the report could trigger one of the identified health problems: severe anxiety.

Despite Subra's failure to prove why Dish residents had health problems, some of them did have problems with headaches, nausea, nosebleeds, and other ailments that might be caused by air pollution. Where was the TCEQ in all of this? In August and October, the commission did some of its own air monitoring at Dish, monitoring sixty sites, some at Dish and some at a distance from town. At one site several miles west of town, TCEQ did find an unacceptable amount of benzene and other chemicals leaking from a compressor station. It also got its own toxicologists to examine the Wolf Eagle report, which, once again, highlighted defects in Wolf Eagle's methodology. The TCEQ's determination of short- and long-term ESLs involved measuring air during one-hour intervals and longer-term measurements taken over the course of a year. Wolf Eagle's measurement of a twenty-four-hour airflow did not permit an effective comparison of chemical levels in either category. To estimate conservatively what a one-hour test might have shown, the TCEQ multiplied those levels by twenty-four, treating a day's exposure as though it were the level for one hour. Even at that, the TCEQ failed to find concentrations of chemicals at levels clearly threatening health. It was possible, however, that some individuals might have suffered adverse effects from low levels of exposure or that chemicals might be present at levels suggesting long-term risk. Wolf Eagle's work did not establish either possibility. The TCEQ toxicologists strongly recommended additional air sampling at Dish.[33]

If the TCEQ thought its response to Dish residents would end controversy, it was certainly mistaken. Regional media played up the perils of exposure to toxic chemicals found by Wolf Eagle. In particular, *Denton Record-Chronicle* reporter Peggy Heinkel-Wolfe, who had recently featured environmental problems created by gas drilling, told her readers that Dish air monitoring showed "emissions at some eye-popping levels." She quoted Subra's verdict on the chemicals: "They can overtake you and make people very sick," a statement that left the issue of levels of exposure untouched.[34] Longtime critic of industry urban operations, the *Fort Worth Weekly*'s Peter Gorman featured Dish in an article titled "Sacrificed to Shale." *The Shorthorn*, campus newspaper for the University of Texas at Arlington, told its readers that "the benzene concentration levels in Dish . . . could cause adverse health effects, including leukemia."[35] In the light of media coverage, the TCEQ seemed at best to be falling down on the job, and at worst covering up for industry environmental damage. Worse yet, when Dish residents met with TCEQ staff on December 14, TCEQ chief engineer for air quality Susana Hildebrand told an angry audience that, though the Commission wanted to do more air testing, it was short of cash—not the sort of answer to soothe tempers. In fact, some residents, like Mayor Tillman, had already made up their minds that they would get no help from state agencies. As he told the *Fort Worth Weekly*, "This is squarely on the shoulders of the TCEQ, and they let this happen to us. I just feel betrayed by the state."[36] Nonetheless, in October, he had asked the Texas Department of State Health Services (DSHS) to test people in Dish for toxic chemicals identified by air sampling. The department did so.

The news from Dish offered ample grounds for alarm back in Fort Worth, particularly as it echoed air pollution charges made by Deborah Rogers a few months earlier. What if gas wells and compressors in town were releasing a toxic cocktail of chemicals into the air in neighborhoods? The TCEQ itself had found benzene when it monitored some sites in Wise and Denton Counties, yet it did not seem to be nearly vigorous enough in determining whether there was a general danger from this known carcinogen, leading local politicians to demand an explanation from the commission. State Senator Wendy Davis called for the lieutenant governor to launch a study of the environmental impacts of gas production in the Barnett Shale. State Representative Lon Burnam went much further. He not only demanded the TCEQ answer nine questions he posed on what it had done about regional air pollution but called for a one-year city and state moratorium on new well-drilling permits so the state could establish the safety of drilling; thus, he challenged not only the TCEQ but also the Texas Railroad Commission. In December, the Fort Worth City Council called on TCEQ engineer Keith Sheedy to tell it what TCEQ air testing in and around the city showed about levels of chemicals in the air; as Mayor Mike

Moncrief put it, "Give us some hard information, some hard numbers." Unfortunately for TCEQ credibility, Sheedy could not offer the council specific data and temporized by saying he would pass along the request to his executive director.[37] What everyone wanted to know was whether lethal quantities of benzene were in the air.

Caught in a firestorm of negative publicity, the TCEQ belatedly decided that it would have to respond more aggressively to public concerns. Castigated by Rogers and Tillman for lack of prompt response to their complaints, the Commission announced on December 17 that it would follow up on any oil- and gas-related complaints within twelve hours. With a view to damage control after its embarrassing grilling by the Fort Worth City Council, TCEQ immediately began an all-out air monitoring within city limits, visiting 126 drilling, compression, and disposal sites. A month later, TCEQ Deputy Director John Sadlier told the City Council that results of that monitoring showed no evidence of toxic chemicals above screening levels. Environmental activists received this news with skepticism—as did the City Council, which decided to hire its own air quality survey from a Massachusetts firm, the Eastern Research Associates; the skepticism proved warranted. In late May 2010, television station WFAA broke the news that the TCEQ had done its December testing with a mobile unit whose equipment was not sensitive enough to pick up levels of chemicals like benzene, dangerous with long-term exposure. With more intensive screening in January, the TCEQ found benzene at four sites. The TCEQ had not shared these findings with the public nor had it passed on findings from natural gas facilities in Wise and Denton Counties; they showed toxic emissions at one out of every five sites. State Representative Marc Veasey called the TCEQ's actions "unacceptable." Davis said the TCEQ continued "to hide data crucial to the public's health." And up in Dish, Tillman said what many in the region were probably concluding about the TCEQ: "These guys are just straight up lying to the public."[38]

While the TCEQ took the brunt of public anger, the Railroad Commission faced its own crisis of credibility. In fact, the Railroad Commission was no better prepared than the TCEQ to pursue inspection of thousands of new gas wells and new miles of pipelines. Not only was the RRC understaffed, having only thirty-two inspectors for the huge area covered by the Barnett Shale, but the Shale sprawled over parts of three RRC administrative districts with offices in Kilgore, Wichita Falls, and Abilene. Tarrant County, for example, fell in District 5, with offices in Kilgore; Denton County was regulated from Wichita Falls. In September 2008, the Railroad Commission finally opened an office in downtown Fort Worth; by that time, the boom was on its way to bust. Like the TCEQ, the RRC was vulnerable to the charge that it had done little to crack down on environmental violations. In 2009, the RRC had identified some 80,000 regulatory violations but issued only 379 penalties, and the maximum fine was only $10,000 a day. This negligence had gotten the attention of the Texas Sunset Advisory Commission, a panel consisting largely of state legislators, which evaluated the functions of each state agency once every twelve years. Unfortunately for the RRC and the TCEQ, 2010 was the year for evaluation. In January 2011, the Texas Sunset Advisory Commission recommended replacing the RRC with an appointed state regulatory agency with redefined functions.[39]

Though there was good reason to question how well both the Texas Commission on Environmental Quality and Railroad Commission of Texas performed, for state officials, there was also good reason to try to protect both agencies—the threat of federal intervention from the EPA. From the 1920s onward, Texas resisted Washington's occasional efforts to intervene in state petroleum industry regulation. In 1935, Texas and other oil-producing states organized the Interstate Oil and Gas Compact Commission as a way of derailing a New Deal push for federal industry regulation.[40] Recently, state oil regulatory agencies formed the Groundwater Protection Council with a view to keeping environmental regulation relating to water in state hands; it argued that state agencies' staffs were more knowledgeable about the geology and hydrology of their regions than federal bureaucrats. But the election of Barack Obama meant a much more aggressive EPA, one prepared to listen to environmental activists' complaints. In November 2009, much to the delight of Barnett Shale environmentalists, the EPA appointed

none other than Al Armendariz as regional administrator, the top environmental official in Texas. Armendariz's record of attacking industry air polluters made him, as the *Fort Worth Weekly* put it, "something of a rock star in environmental circles," and there could be no doubt of his sympathy for their grievances. He let journalists know, "Very soon, maybe 18 months, maybe 24 months, there will be a vigorous air emission program going in Texas."[41]

Emissions from petroleum industry operations, however, was not the only issue the EPA intended to pursue. Heeding environmentalists, it was ready to take on fracking, and in particular, the technology's impact on water. Water pollution would displace emissions as the environmentalists' anti-gas industry focus in 2010, both in the Barnett Shale and elsewhere.

CHAPTER 7
Fear of Fracking

FRACKING—USING MILLIONS OF GALLONS OF WATER IN EACH well—was, of course, essential in unlocking gas and oil from very tight rock formations like the Barnett Shale. It was not surprising, then, that in any region where drinkable water was not superabundant and drought frequent, fracking could be seen as a threat to water supply. But it was water contamination, rather than water supply, that came to dominate national discussion of the technology. Environmental activists have asserted that fracking pollutes water wells and causes aquifer damage, that it sends toxic chemicals into domestic drinking water and methane into home water systems. Because most Americans know little or nothing about fracking, such charges are both credible and horrifying; they would seem to mandate vigorous governmental intervention and industry regulation. But should such regulation be in the hands of federal or state regulators? And is fracking really the environmental hazard its critics say it is? These questions became important as gas drilling continued in the Barnett Shale and advanced into the Marcellus Shale in the Northeast. Both literally and figuratively, the new technology came to be on trial—and it still is.

Country residents in Parker County, largely dependent on their own wells for water, took the lead in making gas drilling's impact on water a Barnett Shale environmental issue. Their initial concern was water supply. Unlike landowners in the cities of Fort Worth and Dallas, which drew much of their water from nearby lakes, Parker County landowners usually tapped the Paluxy and Trinity aquifers for groundwater. Even without a drilling boom, population growth in the county put increasing strain on water resources. Starting in 2005, however, worsening drought diminished the amount of rainfall that could recharge water levels in the aquifer. As underground water levels fell, landowners began to see wells run dry. Small wonder, then, that they started to question industry use of well stimulation technology, which could require up to four million gallons of water to bring gas into a well. Once used in a frack, the water that came back to the surface was laced with frack chemicals, brine, and condensate or oil (or both): it was undrinkable and hauled away for disposal. Would fracking gas wells ultimately leave places like Parker County dry? Would that possibility mean fracking should be banned? In 2006, the Texas Water Development Board hired a study of water use in the Barnett Shale region, which found that fracking was responsible for, at most, only three percent of water use.[1] But even if fracking placed less strain on water supply than, say, lawn watering, a landowner could still wonder if the gas well on the other side of the road was the reason his well ran dry.

In 2008, the focus of criticism directed at fracking shifted from water supply to water quality, and horror stories about water pollution in country wells captured regional media attention. Several landowners in Hill County, for example, said that their water wells began producing brown, foul-smelling water containing sulfates and toluene directly after Williams Production-Gulf Coast drilled two gas wells near them; a hydrologist told one of the landowners not to use the water—"not for showering, cooking, feeding animals, not even for watering the grass." A Johnson County housewife noticed that her well water had an odd smell after a well was drilled near her home, and when she washed her hair, the water turned her hair orange. Tarrant County Commissioner J. D. Johnson said his well water began to turn everything it touched a dark gold color after gas wells near his home were fracked in 2005. And back in Dish, one resident said gas drilling left his well water full of sediment.[2] In these and similar instances, environmental activists singled out fracking for blame.

For those knowledgeable about petroleum industry technology, the possibility that fracking was responsible for contaminated water wells was most unlikely. In the Barnett Shale, seven thousand feet of solid rock commonly separated a fracked wellbore from a water well usually no deeper than several hundred feet. It would be physically impossible for frack water to percolate up through over a mile of rock to a water well. As a method of well stimulation, moreover, fracking had been in use for over fifty years. During that time, there were no reported instances of fracking harming water. But even industry advocates had to admit that, while fracking itself was not at fault, drilling mishaps and irresponsible operation could cause damage. Cementing or casing failures might let frack water escape from a wellbore; a gas blowout at a shallow depth or a nearby well improperly plugged and abandoned might cause problems; drilling waste or used frack water spilled or improperly handled in disposal might contaminate soil and water. To a landowner whose water became unusable, however, the easiest way to explain contamination was to blame fracking. The well water was clean before the nearby gas well was fracked, but afterward, it was dirty. And for many environmentalists, fracking was just one more evil in an environmentally destructive industry.

When it came to water problems, one of the most outspoken Barnett Shale industry critics was Wise County environmentalist Sharon Wilson, and, with Calvin Tillman, she would work to give local problems national attention. Wilson was one of many country dwellers who looked back to life before the boom: "I used to think I'm the luckiest person in the world to live out here. . . . It looked like something out of Norman Rockwell. I wanted a simple life." The shale boom, however, motivated her to take on an activist's mission, to publicize industry irresponsibility and environmental damage by setting up a blog, *Bluedaze*, and by commenting on articles in the *Fort Worth Star-Telegram* as TXsharon. She found plenty to criticize in Wise County alone—oil waste discharge from a sludge pit into a nearby creek, a portable toilet dumping waste beside another creek, neighbors with contaminated water wells, compressor stations emitting clouds of fumes. She told readers of the *Wise County Messenger* in 2008, "Over time, the beautiful rolling hills of Wise

County have become cloaked in a smoky veil." Like Deborah Rogers and Calvin Tillman, she found that complaints to state agencies brought no effective action. So she turned to the national environmental organization Earthworks, and in March 2010, she organized a Texas Chapter of its Oil and Gas Accountability Project (OGAP).[3] It would push for more EPA regulation of gas drilling.

Wilson and her fellow OGAP members could expect a sympathetic response from the EPA because during the previous year, in June 2009, fracking came under fire in Washington. Representatives Diana DeGette (D-Colo.), Maurice Hinchey (D-NY), and Jared Polis (D-Colo.) wished to amend the Energy Policy Act of 2005 to require oil and gas companies to disclose what chemicals they used in frack water. The 2005 act exempted fracking from the act's regulation, in what environmental activists called "the Halliburton loophole," referring to Vice President Cheney's former connection with the oil field service company Halliburton.[4] Their bill, which proposed a Fracturing Responsibility and Awareness of Chemicals (FRAC) Act, did not get through the Congressional session of 2009, but it was reintroduced in the following session. In February 2010, the House Energy and Commerce Committee, chaired by Representative Henry Waxman (D-Calif.) launched its investigation of fracking and directed the EPA to investigate the technology, particularly focusing on hazards to water quality and public health. In fact, the EPA had studied fracking in 2004, concluding it posed "little or no threat" to drinking water, but environmentalists said that study was too limited in scope to be valid.[5] Now the agency had a mandate to do much more. One of the areas it singled out for close study was the Barnett Shale—specifically, Wise and Denton Counties—Sharon Wilson's home turf and Al Armendariz's EPA jurisdiction.[6]

Armendariz was ready to take on the EPA's investigational assignment. Meeting with Dish residents in May 2010, Armendariz reassured them he was bringing the head of the EPA's enforcement division to tour North Texas and reminded them that "the EPA is not toothless." He emphasized the EPA would study fracking, observing, "I would be very surprised if Dish, and the Barnett Shale in general, isn't going to be part of that study."[7] In terms of EPA attention, Armendariz proved as good as his word. On July 8, the EPA held a public meeting on fracking at the downtown Fort Worth Hilton, drawing a capacity crowd of six hundred persons. In the course of the meeting, Mayor Calvin Tillman held up a container of dirty water from a Dish water well, calling for focus on what fracking did to drinking water. Linda Hanratty, of the League of Women Voters of Tarrant County, called for the EPA to make a case study of the Barnett Shale gas development. Sharon Wilson put it more directly: "I'm sending out an SOS to the EPA. We need you here. We need you on the ground. We need you now"—a plea which drew vigorous applause. By contrast, when Railroad Commission Chairman Victor Carrillo tried to defend fracking, as well as RRC turf, by noting, "There are no documented cases of fracturing causing groundwater contamination in Texas," he drew boos from some in the crowd. That led meeting moderator Adam Saslow to remind the audience to use "manners your mother taught you."[8] Apparently many in the crowd had already made up their minds, both on fracking and on the Railroad Commission.

Two weeks before the EPA held its Fort Worth meeting, some audience members may have made up their minds about fracking with the help of an HBO airing of Josh Fox's *Gasland*, a documentary exposé of the evils of gas drilling in general and fracking in particular. Not only could North Texas residents watch *Gasland* at home, but on June 21, Representative Lon Burnam's supporters invited everyone to a *Gasland* watch party. For that matter, Texas OGAP had shown it at a fundraiser in May.[9] As a means to terrify viewers about the dangers of fracking, *Gasland* was highly effective, if anything but subtle. Travelling from gas fields in northeastern Pennsylvania to those in Colorado, Wyoming, and North Texas, Fox filmed environmental horrors: contaminated well water in Pennsylvania, sick people and animals in Wyoming, water from a household faucet ignited into a fireball in Colorado. When he got to Fort Worth, Fox stayed with Don Young while he talked with Calvin Tillman, Al Armendariz (not yet appointed regional EPA head), and Wilma Subra. Once *Gasland* was released, it received fervent applause from both film and industry critics. It won a special jury prize at the Sundance Film Festival. The *New York Daily News* decided it showed why New York state should make banning fracking a "top priority." The *New York Times* found fault with Fox's filming but told readers, "If you are predisposed to distrust big business and the bureaucrats who regulate it, then *Gasland* . . . will light a flame in you."[10] As anti-industry ammunition, *Gasland* was everything an ardent environmentalist could hope for.

Apart from showing a novelty like igniting water from a kitchen sink faucet, one reason *Gasland* was as effective as it was in rousing public fear of fracking was that very few Americans outside the oil patch, let alone those within it, knew what fracking was. If it could break up rock to release natural gas, why wouldn't that gas end up in people's water wells? Fox filmed places most viewers had never heard of—Dimock, Pennsylvania; Pavillion, Wyoming; Dish, Texas. Who knew whether what they saw and heard was misleading? Fox juxtaposed drilling rigs and flaming faucets—he did not tell his audience that, in many parts of the country, seeps from shallow deposits of oil and gas, seeps present before any oil or gas drilling, commonly made their way into water wells, springs, and streams. Nor did he mention that, in some Pennsylvania localities, runoff from abandoned coal mines had polluted water before gas drilling took place. *Gasland*'s visual images were far more powerful than the industry disclaimers they provoked. After all, people knew what they saw—didn't they? And when it came to believing that the petroleum industry was guilty of environmental damage in the summer of 2010, one had only to watch the nightly news for additional confirmation of that belief: evidence in the form of the great BP Macondo blowout in the Gulf of Mexico.

The time was right for a *Gasland* event in North Texas—one allowing the EPA to flex regulatory muscle, ideally on fracking-related environmental damage. One such occurrence emerged in August at two homes in southern Parker County, one of which was owned by Steven and Shyla Lipsky. The Lipskys would take the lead in bringing about EPA action.

Early in August 2010, Steven Lipsky complained to the Railroad Commission's regional office in Abilene about gas in his water well. In March and June 2009, Range Resources had drilled two gas wells just over the county line in Hood County. Both wellbores were within several hundred feet of property owners' water wells in terms of horizontal location, but over a mile beneath them. The Railroad Commission collected water samples from Lipsky's well on August 17 and gas samples on August 26. It also required Range to test gas from its wells' bradenheads and tubing to see if their gas was chemically identical to that in Lipsky's water.[11] But that still left Lipsky with gassy water and no apparent remedy from the RRC.

Even before the Railroad Commission tested his water well, Lipsky was in contact with Alisa Rich of Wolf Eagle Environmental, though exactly how he found her is unclear; he said he had talked to "a woman . . . but I can't remember her name . . . I think she's in Texas, though."[12] In any event, Rich urged him to do air tests near his water well—if there was gas in the air, he could ask for help from the TCEQ and then the EPA. Rich then tested both water and air on Lipsky's property. As at Deborah Rogers's farm and at Dish, she used a twenty-four hour air sample and identified quantities of methane and other chemicals. As before, her findings were alarming.[13] Here was ammunition to offer Armendariz and the EPA. Better yet, Lipsky invited Parker County Judge Mark Riley to his home, showed him what was apparently a water hose, put a lighter to the nozzle, and let him watch as the hose flamed, persuading him there was a problem. Riley observed, "I'm certainly not a scientist, but anybody with common sense would know an issue existed." What Riley did not know was that the hose was attached to the water well to vent its gas.[14] *Gasland* had come to Parker County.

In October, the EPA took its own samples of gas and water from the Lipsky property, as well as from Range's wells. Several days earlier, the Railroad Commission had monitored testing of production casing and tubing on both gas wells and found no flaws in either that could have led to leaks of gas; its analysis of gas from the Range wells and the Lipsky well was in process. At the end of November, the EPA told Range that the gas in Lipsky's well was the same as the gas from Range's wells, notwithstanding the reservations of at least one staff scientist who thought the data too limited to arrive at that conclusion. The EPA did not try to explain how gas travelled from Range's wells to Lipsky's well, nor did it consider what it said it knew: that natural gas seeps were common in area water wells.[15] No matter: Al Armendariz was ready to move. Late in the afternoon of December 7, he told Sharon Wilson and other friends that he was in the process of releasing news that the EPA was moving against Range. As he put it, "We're about to make a lot of news . . . time to TiVo channel 8." He added, "Makes me think about . . . when I first appreciated the magnitude of poor fluid management practices from pictures and video on Sharon's blog." Sharon replied, "Yee haw! Hats off to the new Sheriff and his deputies!"[16]

Having alerted his friends, Armendariz let the news of the EPA's emergency order against Range reach Fort Worth's WFAA News 8, explaining he did so after the Railroad Commission declined to act in the case. Only after giving the press release to the media did he have EPA staff tell the Railroad Commission what his agency ordered. Range was to bring drinking water to the homeowners, give them gas monitors for their homes, sample soil and all nearby water wells for gas, develop a plan for aquifer remediation, and check the equipment of its nearby wells. There was, the EPA stressed, danger of "fire or explosion."[17] Primed for a showdown on the turf of federal versus state regulation, Armendariz had thrown down the federal gauntlet to the Railroad Commission without letting courtesy get in the way. He also gave journalists an irresistible story, which they were quick to exploit: the *Star-Telegram* ran an editorial with the memorable title, "The Texas Railroad Commission Fiddles While Well Water Burns."[18]

Stung by the EPA's preemptive regulatory strike, the Railroad Commission was ready to give as good as it got. RRC Chairman Victor Carrillo told the press, "This EPA action is unprecedented in Texas." Commissioner Elizabeth Ames Jones declared the order was "another EPA action designed to reach pre-determined conclusions and to generate headlines rather than conduct a successful environmental investigation." Commissioner Michael Williams put the spotlight on familiar states' rights turf by saying the EPA was engaging in "Washington politics of the worst kind," explaining that "the EPA's act is nothing more than grandstanding in an effort to interject the federal government into Texas business." But however forceful the Railroad Commission's rhetoric, the fact remained that it had not concluded what the source of gas in Lipsky's water well was. Its stand was readily challenged by Armendariz, who assured journalists, "We were worried about the families' safety. It was incumbent for us to act quickly."[19] As it was for the Railroad Commission: it called a hearing on January 10.

Once the sensation caused by the EPA's order died down, others outside the Railroad Commission's ranks began to question the EPA's charges against Range Resources. On December 13, Gene Powell, publisher/editor of the *Powell Barnett Shale Newsletter*, issued an eight-page article challenging the EPA decision that Range contaminated the water wells near its gas wells. It cited veteran Parker County water well drillers, among them Larry Peck, who said he had drilled a water well near the Lipskys in October 2005, long before local Barnett Shale gas drilling, and brought up enough gas for a five-to-six-foot flare. Peck offered to plug the well, but the owner said he would use the gas![20] Other water-well drillers agreed that gas in local water wells was very common, coming from the Strawn Formation below the water-bearing Paluxy Formation. Drill as deep as two hundred feet and one was likely to hit the Strawn. For that matter, as more and more water wells tapped the Paluxy for new homes, the water level in the Paluxy dropped—bringing its level nearer to the Strawn gas. Falling water levels might explain why Lipsky's well could suddenly start producing gas. There was also the possibility that long-abandoned shallow gas wells might offer a pathway for gas to enter water. In short, one could not immediately blame Range. But such considerations did not sway the position of EPA defenders like Sharon Wilson. She told the *Star-Telegram*, "If my home were in danger of exploding and I could shoot fire from my garden hose like a flame thrower, then I, and I would wager most Texans, would be grateful for any agency at any level that stepped in to protect me."[21]

Range Resources, meanwhile, had been working with the Railroad Commission to try to determine if its wells were responsible for water contamination, and it tried to work with the EPA, meeting with the agency's staff on December 15. At that meeting, EPA staff agreed with Range representatives that fracking had not caused gas in the water wells and acknowledged that water in the area often contained gas, even before gas drilling. The staff, moreover, could not explain the means by which gas from Range's wells could be reaching the water wells. But if Range thought its arguments would mean the EPA would withdraw its order, the company soon found otherwise: the next day the EPA told Range it would not back down. Additionally, it gave Range the task of showing how gas from its wells migrated into groundwater. When Range could not show how its wells had contaminated area water, the EPA said Range violated its order and was subject to a penalty of $16,500 a day. Range asked to know what findings the EPA had to justify its order. The agency stonewalled in response, refusing to share information or let Range question its staff. Range then obtained a court order allowing it to depose John Blevins, the EPA Regional Director of Compliance Assurance and Enforcement.[22] His testimony showed the agency had good reason to stonewall.

What Blevins admitted amounted to a demonstration of how the EPA had rushed to judgment. It had not considered the possibility that naturally occurring Strawn gas was contaminating water wells; in fact, it had not evaluated the geology below the Lipsky water well at all. It had not analyzed Strawn gas on the chance its composition would match gas in Lipsky's well. It found the hydrocarbons in the gas from Lipsky's well were like hydrocarbons in Range's gas—but failed to compare the nitrogen content of the two, which would turn out to be markedly different. Blevins would only say Range might have contaminated the wells, not that it did, and he did not know whether fracking played any part in what was at issue. In effect, given the slipshod nature of the EPA's investigation, the agency assumed Range was guilty unless it could prove it was innocent. By insisting Range show how its gas migrated to the Lipsky well, the EPA was asking Range to find something its own staff could not show was there.[23]

On January 19, the Railroad Commission began hearings on the Range case, having delayed several days to permit deposing Steve Lipsky and Alisa Rich. Neither appeared at the hearing, nor did anyone appear from the EPA, though a bevy of geologists, engineers, and water-well drillers testified in support of Range. A key element in evidence produced at the hearing involved comparing the chemical composition of the gas in water wells to the Barnett Shale gas from Range's wells. Here Mark McCaffrey of Weatherford Laboratories, Houston, pointed out that the EPA's gas analysis was flawed because it failed to look at the nitrogen content of gas samples. The gas from the Lipsky and other water wells was high-nitrogen Strawn gas: it was not the same as what Range's wells produced.[24]

Even if the gas Range produced was not the same as gas in water wells, could fracking have somehow created faults through which gas entered those wells? Norman Warpinski of Pinnacle—a Halliburton Service—testified that data from 320 frackings showed fracture lengths of no more than four hundred feet; over four thousand feet separated wellbores being fracked from water formations tapped by domestic wells, making it impossible for a fracture to reach as far as water wellbores. Alternatively, could natural rock fractures let gas escape from the Barnett to the water wells? Geologist Charles Kreitler testified that seismic data showed no such faulting in the area. What about the allegation by Alisa Rich and the EPA that gas levels in the air on the affected properties were so high there was a risk of fire or explosion? Keith Wheeler testified that of air samples taken, the highest concentrations of hydrocarbons were 6 parts per million (ppm) for ethane, 13.9 ppm for methane, and 61 ppm for propane. The lower explosive limit for ethane was 30,000 ppm, for methane 50,000 ppm, and for propane 21,000 ppm. The highest concentration of these gasses in 117 soil samples was less than 0.2 percent of the lowest explosive limit for each.[25] Perhaps that explained why no one's house exploded. In any event, on March 22, after two months' examination of evidence, the Railroad Commission ruled that Range had not contaminated nearby water wells.

The Railroad Commission's decision for Range did little to resolve controversy. Not surprisingly, since it challenged the Railroad Commission's authority, the EPA did not back away from its order to Range. Instead, it hired another analyst, Geoffrey Thyne, to analyze water samples from the area. Thyne decided the gas in well water could have originated in Range's wells. Under the guidance of Armendariz, moreover, the EPA in Texas escalated action on air pollution, attacking coal-fired electric plants for toxic emissions of mercury, lead, and arsenic. Armendariz also brought Wise and Hood Counties into the North Texas area covered by the federally mandated anti-smog program, prompting an attack from Texas Governor Rick Perry who called the move "trampling on Texas rights."[26]

Similarly, the Railroad Commission's decision did nothing to deter litigation. Range sued the EPA in federal court. US District Judge Royal Fergeson would not dismiss the EPA order, but blocked levying the financial penalties set by the federal agency. In June, Fergeson also postponed further action in the case until the US Fifth Circuit Court of Appeals ruled on a Range complaint relating to EPA procedures. The same month, the Lipskys filed a $6.5 million lawsuit against Range in State Forty-third District Court. In turn, Range filed a $4.2 million suit against the Lipskys for defamation. Judge Trey Loftin ruled that Alisa Rich conspired with Steven Lipsky to defame Range by creating the video showing flame coming from the hose connected to Lipsky's water well. In fact, this and other environmental litigation brought Rich's credentials as a scientific analyst under harsh scrutiny. Critics pointed out that Rich had to change her firm's name because she was not an engineer, nor did she employ any engineers. Though she said she had a PhD and consulted with petroleum companies, she was forced to admit she was only enrolled in the PhD program at the University of Texas at Arlington and had never in fact been hired as an oil and gas consultant. When the *Fort Worth Star-Telegram* asked Armendariz to comment on Rich's work, he declined. In another environmental suit, State District Judge Tom Lowe refused to accept Rich as an expert witness.[27] But if Rich's credibility was doubtful, where did that leave those who used her findings to attack gas drilling?

Because the Range case, in effect, put fracking on trial, its proceedings got extensive publicity for the EPA agenda and spurred national reaction, though that was not entirely what the EPA might have wished. In Washington, Senator James Inhofe (R-Okla.), a stalwart petroleum industry supporter, demanded the EPA deliver documents relating to Armendariz's December 7 order. The *New York Times* told its readers how Armendariz alerted his friends and the media about his order before he let the Railroad Commission know about it, and quoted Sharon Wilson's exultant "Yee haw" response. It also reported that Alisa Rich "strategized" with Lipsky to move investigation of his complaint from the Railroad Commission to the EPA. In May, Elizabeth Ames Jones told the US House Committee on Science, Space, and Technology that assertions that fracking polluted groundwater were "fairy tales." She cited the Range case as evidence that the EPA was unfit to regulate gas drilling.[28] The EPA, however, continued working to extend its regulation to gas drilling and fracking. In July 2011, it proposed a set of regulations designed to limit emissions of volatile organic compounds from natural gas present in well completions, in addition to chemicals released from fracking flow-back fluid, compressor stations, storage tanks, and other sources. In December, it issued a draft of its findings on Pavillion, Wyoming, water contamination, suggesting chemicals in water were probably the result of fracking. Meanwhile, in May, Energy Secretary Steven Chu assembled a shale gas advisory panel drawn from members of his advisory board and outside experts to investigate the environmental impacts of shale gas production.[29] Having launched their probe of gas drilling operations, Washington regulators seemed to be moving ahead at full speed, which industry critics could applaud.

Much to the dismay of industry critics in North Texas, the events of 2012 would bring two major setbacks to the environmental cause. In March, the EPA dropped its order and court action against Range Resources and set aside Thyne's report. Offering scant explanation for its change of heart, it simply said it would focus on "the science and safety of energy extraction." It did not bother to notify Steven Lipsky, who learned of its decision from a reporter. Outraged, regional and national environmentalists would ask for investigation of the agency's decision the following year; Earthworks Energy Program Director Bruce Baizel observed, "Range Resources went behind closed doors and solicited special favors."[30] But for North Texas environmentalists, much worse was to come.

When Al Armendariz met with Dish residents in May 2010, reassuring them that the EPA would act on their complaints, he also outlined what his approach to regulating oil and gas operations would be, which was captured on video by someone at the meeting. As he put his "philosophy of enforcement":

> It was kind of like how the Romans used to conquer little villages in the Mediterranean. They'd go into a little Turkish town somewhere, they'd find the first five guys they saw, and they would crucify them. And then you know that town was really easy to manage for the next few years. And so you make examples out of people who are in this case not compliant with the law. . . . You hit them as hard as you can and you make examples out of them.[31]

These remarks, apart from showing Armendariz a better engineer than ancient historian, were splendid ammunition for industry advocates who argued the EPA was out to conduct a witch hunt against the petroleum industry, as well as for Congressional Republicans out to show President Barack Obama pursuing an environmental agenda. The video, shown on YouTube, proved to be just what Senator James M. Inhofe needed to deliver a broadside against the EPA. On April 25, 2012, in a lengthy letter, Inhofe asked EPA head Lisa Jackson to explain the EPA's action against Range Resources and Armendariz's comments at Dish. Inhofe charged the EPA had a "proactive interest in the federal regulation of hydraulic fracturing," an interest that "contributed to the agency's preemptive actions." Citing the EPA's statements implying fracking contaminated water not only in Parker County, but also Pavillion, Wyoming, and Dimock, Pennsylvania, he accused the EPA of acting on the basis of "preconceived conclusions" and "political activism."[32] For the EPA, it was bad enough to have an angry senator at its door, but its regional administrator was responsible for it. Armendariz's remarks were as highly embarrassing as they were eminently quotable. Perhaps the EPA needed to make an example of Al Armendariz.

In Washington, the EPA's immediate reaction to Inhofe's attack was an attempt to distance itself from its indiscreet administrator. Despite a hurried apology from Armendariz for "poor choice of words," EPA head Lisa Jackson hastened to assure the press that his remarks were "disappointing" and "not representative" of EPA policy. The White House, well into a presidential reelection campaign, was ready with disclaimers as well. When questioned by a reporter from Fox News, Presidential Press Secretary Jay Carney echoed Jackson by observing that Armendariz's perspective on regulation was "clearly not representative of . . . the President's belief in the way that we should approach these matters." But Armendariz was far too quotable for such damage control to work, as Inhofe knew. Broadening his attack to include the White House, he asserted, "There are Armendarizes all throughout this administration," and used the sound bite, "crucifixion philosophy." *Wall Street Journal* columnist Kimberley Strassel took this idea and enlarged upon it with a feature on Armendariz and the Range case called, "The 'Crucify Them' Presidency." With both the EPA and the White House at bay, Republican Congressmen swooped in for the kill: twenty-nine Representatives from Armendariz's Region Six, as well as Congressmen from Iowa and Arizona, clamored for the administrator to resign.[33] An outstanding liability to both the agency and the administration, there was little chance Armendariz could keep his job.

And so Armendariz sent his resignation to Lisa Jackson, effective April 30. Not surprisingly, his departure was applauded by the TCEQ, which commented, "Dr. Armendariz's mistake was that he slipped and unveiled the EPA's questionable and draconian enforcement philosophy." The Railroad Commission's Barry Smitherman called for an investigation of Armendariz to determine "how many times he crossed the line and harmed our economy and our energy future by pursuing his

extreme political agenda instead of science and fact." US Representative Michael Burgess (R-Tex., Twenty-sixth District) called Armendariz's approach to regulation "a practiced pattern of hostility." But for those sympathetic with the environmentalist position, Armendariz's resignation was dismaying. The *Austin American-Statesman* editorialized that, while he made "the right decision to resign," he was the victim of "a fabricated campaign." In her *Bluedaze* blog, Sharon Wilson told readers, "Big Gas Mafia wins again, our beloved Dr. Al has resigned. . . . Drilling impacted communities lost a champion in the fight to improve the fracking industry's lamentable track record of sacrificing community health and clean water for the sake of maximizing corporate profits." Seeing Armendariz as one of the best Region Six administrators, the Sierra Club's Ken Kramer called his resignation "a major loss for Texas." It was a gain for the Sierra Club, however, which hired Armendariz as a lobbyist in September.[34]

While the whirlwind of controversy over federal regulation targeting fracking played on, some Texas legislators introduced a variety of bills aimed at tightening state regulation of shale drilling. In 2011, Lon Burnam, who had long acted on his environmental convictions, introduced eight bills relating to gas drilling, including one to require "green completions"—completing wells using vapor recovery systems that limited emissions, and mandating a 1,200-foot setback between gas wells and public school boundaries. With Representative Jim Keffer (R-Tex., Sixtieth District), he was part of a bipartisan effort to require disclosure of chemicals used in fracking. This measure would perhaps head off the kind of regulation by Washington in the proposed FRAC Act, since Texas could say it already required disclosure. Because the measure that eventually passed allowed exemption of chemicals that were proprietary secrets from disclosure, however, environmentalists criticized it as inadequate. Topping Burnam for numbers, State Senator Wendy Davis introduced fourteen gas-related bills, including one requiring addition of a chemical "tracer" to frack fluid in order to determine whether it went from wells to contaminate water, as well as her own frack fluid disclosure bill. Environmentalists like Sharon

Wilson applauded such legislative efforts, but little came of them, which did not surprise Wilson and other petroleum industry critics. Wilson summed up a disappointing session by reflecting, "The Texas legislature is controlled by oil and gas interests."[35]

While the Texas legislature failed to deliver measures North Texas environmental advocates favored, they were not about to abandon their campaign against fracking. Even if one disagreed with their conviction that fracturing rock many thousands of feet below domestic water wells would contaminate well water, there were other issues raised by use of the technology, in particular those relating to disposal of the wastewater flowing back from a fractured well. Disposal of wastewater could pose problems ranging from soil and water contamination to mild earthquakes. Here the technology was not at fault as much as those who used it irresponsibly. And, once again, it was country residents who were most at risk from irresponsible industry operations, in part because the City of Fort Worth permitted only one disposal well within city limits. That meant a proliferation of disposal wells in Wise, Denton, and Parker Counties as Tarrant County was drilled up, as well as disposal wells in places like Johnson County, which had been country until the boom.

The disposal of wastewater created by oil field drilling has long been an industry problem, because petroleum drilling and production commonly bring up amounts of very highly mineralized brine from underground. Many times saltier than seawater, the brine is usually laced with some petroleum or volatile hydrocarbons. By the 1970s, oil field brine was handled by disposal in injection wells reaching down to rock formations containing nonpotable water. But the new fracking technology created massive amounts of flowback water. This water might not be as mineralized as oil field brine, but it, too, might contain small amounts of petroleum, volatile hydrocarbons, naturally occurring radioactive materials, frack chemicals—in other words, an unpleasant chemical soup. It had to go somewhere, but there was much more of it needing immediate disposal than had been the case with ordinary oil field brine, which still could be part of the disposal problem in Barnett

drilling.

Needing some way to handle wastewater, Barnett Shale operators usually collected wastewater in tanks near the well and hired disposal companies to truck wastewater from tanks to disposal wells, which might have been at a considerable distance from the well. The tank trucks, carrying wastewater and often weighing some eighty thousand pounds or more, would then travel roads ill-adapted for such heavy loads to the disposal well site. There the wastewater would ultimately be injected downhole into the briny Ellenburger formation, where the water was not drinkable by any stretch of the imagination. But problems could occur between the gas well and the disposal site.

Disposal was not cheap. Companies commonly charged at least sixty cents per barrel to take wastewater for disposal. Companies that charged by the barrel, hence making more money the more barrels they carried, might cut corners by disposing of brine along the highway or dumping wastewater in dry creek beds. Allowing that a disposal company was responsible and took the wastewater to the disposal site, the disposal well had to be cased and cemented properly to ensure that none of the wastewater escaped the wellbore into potable water formations or surface soil. But even if everything was in proper order underground, spills of wastewater could happen before injection. In short, disposal was no easy matter. And if you owned property next to a disposal site, even with the most responsible operation, you would see a constant stream of tanker trucks next to your property—hardly an enhancement to property value.

What no one in the Barnett could have anticipated was that injections of huge amounts of wastewater underground might trigger small earthquakes. The first noticeable tremors took place in September 2008, seven weeks after Chesapeake Energy drilled and started using wastewater disposal wells at the southern end of DFW airport. Between November 2008 and January 2009, seismologists led by Brian Stump and Cliff Frohlich recorded eleven tremors, all too small to be felt by area residents but all clustered in the area of the disposal wells and near an underground northeast-southwest fault. The following June,

Cleburne residents experienced five earthquakes within two days. Like those at DFW, these were too small to do much beyond rattling dishes and windows, but again they were clustered near a disposal well used by Chesapeake Energy. At this point, scientists hesitated to say the disposal wells were responsible for the tremors. There were, after all, upward of two hundred disposal wells in the Barnett Shale, and no tremors had been reported near them. Still, when Chesapeake Energy shut down the suspect disposal wells in August, the quakes stopped, which made it reasonable to believe they may have at least contributed to the tremors.[36]

Though industry advocates were at first skeptical about seeing a connection between disposal wells and earthquakes—there were, after all, thousands of disposal wells in Texas that seemed to produce no problems—there was reason to believe that injections of huge quantities of wastewater into rock in a limited area could generate earthquakes. In 1966, for example, injections of large volumes of water used in chemical weapons manufactured at the Rocky Mountain Arsenal near Denver (closed since 1992) caused over seven hundred quakes within five miles of the disposal well. In Texas, oil and gas production in Lamar and Gregg Counties had been suspected of causing tremors. And since the quakes at DFW and Cleburne, there have been numerous instances of tremors near disposal wells in Arkansas, Ohio, and Oklahoma, as well as a swarm of quakes in 2013 and 2014 near the Parker County towns of Azle and Reno. One common element in these instances seems to be the presence of underground faults in rock near the disposal area, but seismologists are still discussing exactly how that might trigger tremors.[37]

A viable alternative to hauling millions of gallons of wastewater to disposal wells consisted of cleaning the wastewater and reusing it in subsequent fracking. In 2004 the Canadian-based firm Fountain Quail Energy Services began working in the Barnett with Devon Energy, using mobile water recycling units on Devon Energy's leases. Flowback water was turned into a lined "feedwater pit," clarified to remove all solids, and fed into a mobile evaporator unit fueled by natural gas from nearby wells. Heated to steam and then condensed, one hundred barrels of wastewater could yield anywhere from seventy to eighty-five barrels of clean water ready for reuse, leaving the remaining fraction for disposal. Not only did this cut the amount of water to be disposed of dramatically, but it meant far less draw on local surface or groundwater. By October 2010, Fountain Quail Energy Services' site in Wise County was handling over 386,000 gallons of water per day. Unfortunately, this approach to wastewater cost more than three times what it would to haul water to disposal wells.[38] The technology was affordable for a large company like Devon Energy, but far less so for smaller operators challenged by low natural gas prices after 2008.

As the debate over fracking, in particular, and shale gas drilling continued, Charles G. Groat and Thomas W. Grimshaw of the University of Texas at Austin Energy Institute assembled a panel of energy experts headed up by three senior contributors to address some of the controversy generated by new technology and development. In February 2012, the researchers released a 442-page report, the summary of whose findings took 60 pages. Titled *Fact-Based Regulation for Environmental Protection in Shale Gas Development*, the report reviewed some of the main issues relating to shale drilling, looked at media coverage of issues (which was overwhelmingly negative), surveyed environmental impacts, including the question of whether fracking contaminated water, and described existing federal and state regulations and enforcement, all to help policymakers and regulators handle development issues. With respect to water contamination, the experts found no evidence that the actual fracturing process could release chemicals into aquifers; surface spills of frack fluid, however, might contaminate groundwater. They also noted that ordinary oil and gas drilling, apart from fracking, might result in water pollution. As for methane in water wells, the experts saw it as the work of nature rather than man, a phenomenon present before drilling.[39] In short, here was a report that industry advocates could endorse as the definitive scientific verdict on fracking.

Ironically, far from settling differences, the report generated scandal as well as more controversy. In July 2012, it was revealed that Groat sat on the board of directors of an energy company, Plains Exploration & Production Company; that he owned over $1.5 million in Plains stock; and that he received $413,900 in cash and stock from the company in 2011. He was, to all appearances, scarcely a disinterested researcher. Deeply embarrassed, the University of Texas (UT Austin) appointed a review panel to look into conflict of interest. In December the panel released a scathing critique of both the study, which it said had distorted conclusions and fell short of standards for scientific work, and UT Austin's ethics rules for research. Groat retired, and the University of Texas at Austin Energy Institute's executive director, Raymond Orbach, resigned.[40] With its objectivity thus called into question, the report could scarcely serve as a credible guide for policymakers and regulators.

In both North Texas and the nation, then, the controversy over fracking and its impact on water remained unresolved. In North Texas, fear of fracking's depletion of groundwater diminished as Barnett Shale drilling fell off when natural gas prices stayed low. In other parts of Texas, where fracking to produce oil proved dramatically successful and drew on large volumes of groundwater in a time of severe drought, fear of groundwater loss increased. Notwithstanding repeated assurances from geologists and petroleum engineers that fracking posed no threat to domestic water wells, environmentalists and owners of wells producing gassy or mineral-contaminated water continued to blame use of the technology for their problems. As the Range Resources case showed, testimony from a wide variety of experts claiming to use science did little to alter the convictions of either industry proponents or environmental advocates. For industry defenders, the Range Resources case showed how a blameless company was the victim of environmentalists' rush to judgment based on biased and flawed investigative methodology. For environmentalists, it demonstrated how biased state regulators were more interested in cooperating with the industry than protecting either citizens or the environment. Environmental protection would have to come from Washington.

But could environmentalists be sure that help lay in federal hands? Not only were they disappointed when the EPA dropped charges against Range Resources, they were dismayed shortly thereafter when the agency accepted the resignation of their regional champion, Al Armendariz, whose very regulatory zeal had proved embarrassing in Washington. Such reverses seemed to call for even more strenuous efforts not only to regulate and limit fracking but also to ban it altogether. Earthquakes were the least of its hazards. Fracking represented virtually everything environmentalists saw objectionable in oil and gas development. And inasmuch as that very development would delay replacing use of fossil fuels by other energy sources, the technology assuring an abundant future supply of such fuels had to be prohibited. It was at the root of environmental evil.

CONCLUSION
In Retrospect

IN SEPTEMBER 2008, THE NUMBER OF DRILLING RIGS AT WORK in the Barnett Shale peaked at 203. On March 29, 2015, only one rig was drilling in the Barnett.[1] The boom was a memory, but the impact of the opening of the shale upon the petroleum industry, as well as on the region that boomed, is still very much present. So are the issues raised by the gas drilling that opened the Barnett. From the vantage point of nearly two decades after the major Barnett breakthrough by Mitchell Energy and Development, it is worth asking what difference the shale boom made to a region, an industry, and the foreseeable energy future. After all, the history of the petroleum industry is one of booms and busts. Why was this boom different?

Like many other booms, the Barnett Shale boom created a tremendous amount of wealth and economic growth in a very short time. Unlike some booms, however, when boom gave way to bust, that wealth and growth did not disappear. In 2014, the economists of the Perryman Group reported on the economic impact of the boom in North Texas on jobs and tax revenue to state and local government. They found that, notwithstanding the end of the boom, by the end of 2013 it was responsible for creating 107,650 permanent jobs. Regional pipeline activity alone was responsible for nearly 5,300 jobs. Regional oil and gas exploration, drilling, processing, industry supply, and other industry operations supported nearly 94,000 jobs. Since its takeoff in

2001, Barnett Shale action resulted in over $6 billion in tax revenue to the state and $4.5 billion to North Texas cities, counties, and school districts. The big winners in terms of both jobs and tax revenue were the core area counties of Denton, Johnson, Tarrant, and Wise; they saw 84 percent of job growth and 86 percent of tax windfall. Considering the whole North Texas region, the Perryman Group concluded Barnett Shale action was and continues to be a "notable" economic stimulus even after the boom peaked. For that matter, much of this regional economic growth happened when the rest of the nation struggled with economic downturn.[2]

Once natural gas prices sank below $4 an Mcf, Barnett gas drillers had to reexamine strategy. Obviously there was less reason to pick up more leases or rush to get gas into an unpromising market. Some operators cut back on projects, drilling enough wells to hold expensive leases but closing them in, once drilled, in the hope of better prices. Others, like Devon Energy, could have enjoyed healthy revenue from gas wells drilled before the bust, but instead they shifted focus toward processing gas containing natural gas liquids. As part of that strategy, Devon Energy expanded its Bridgeport gas-processing plant and drilled gas wells with longer lateral bores in liquids-rich gas areas. Devon Energy used more frack stages along the laterals and thus produced more liquids, which were more profitable than gas. Devon Energy also had the advantage of picking up oil and gas leases in northern and western Wise County from Mitchell Energy and Development. Since oil prices remained high, Devon Energy could direct project capital to oil as well as gas.[3] Going after oil would prompt Devon Energy to look for it in other shale basins.

Many other operators also shifted focus to parts of the Bend Arch-Fort Worth Basin Province where natural gas was richer in liquids and oil. Wise, Montague, Jack, and Parker Counties were in that category, and operators like Pioneer Natural Resources, which entered the Barnett play relatively late, began drilling for liquids and oil rather than dry gas. But it was Houston-based EOG Resources that had the most success in the oily Barnett. It had leased more than two hundred thousand acres in Montague, Clay, and Archer Counties and got a gratifying amount of oil, as well as liquids-rich gas and condensate. Having mastered technology that could extract oil from shale in North Texas, EOG Resources went on to become a major oil producer in the North Dakota Bakken Shale Play. In the process it went from having 79 percent of its production in natural gas in 2007 to 89 percent of its production in crude oil and other petroleum liquids in 2014. Its strategy yielded handsome profits when oil brought in one hundred dollars a barrel.[4]

Those not positioned to try their luck in the oily Barnett always had the option of finding a buyer for what they had in its dry gas area, selling out, paying bills, and moving on—a time-honored independent oil strategy and one especially appealing as hot new shale plays opened up elsewhere. Among the operators with a leading role in the drilling boom, Range Resources sold most of its Barnett properties in March 2011, having shifted its focus to the Marcellus Shale, where it was one of the earliest drillers. That same year, KKR Natural Resources bought properties from Carrizo Oil & Gas. The following year, Atlas Resource Partners (now Titan Energy) also bought properties from Carrizo Oil & Gas. Pioneer Natural Resources, having decided to move in an oily direction, shifted attention to the Eagle Ford Shale and to the Permian Basin, the latter where it had had its roots, and put Barnett properties on the block. In short, between September 2010 and October 2012, an estimated $4 billion in Barnett assets changed hands.[5]

What of the fortunes of the second-largest—and undoubtedly the highest-flying—of the gas producers active in the Barnett Shale, Chesapeake Energy? As the *Wall Street Journal* reporter Russell Gold pointed out, its CEO Aubrey McClendon pursued a strategy based on the idea that whichever company could grab the most potential gas-producing acreage would succeed. As soon as a shale hotspot seemed to appear, be it in the Haynesville, Fayetteville, Marcellus, Utica, or Barnett Shale, Chesapeake would plunge in, rushing to outbid and outdo competitors for acreage. To keep leases in the Barnett as elsewhere, the company had to drill hundreds of wells each year. In fact, between 2004 and 2011, Chesapeake drilled more wells than any other company in the world!

But to do so, it had to take on a staggering debt load, since long before drilled wells paid out, Chesapeake was paying millions of dollars for new leases. Apparently this did not upset McClendon, who believed that natural gas would be in ever-shorter supply and that, in typical boom-time thinking, natural gas prices had nowhere to go but up.[6]

Unfortunately for both Chesapeake and himself, McClendon was wrong, and when gas prices fell later in 2008, the company and its CEO scrambled to handle debts. Worse yet, Chesapeake shares, like those of other companies, plummeted in the great stock market slide of October 2008, losing some 59 percent of their value. Pressed for cash, Chesapeake not only laid off a large portion of its army of landmen and looked to sell stakes in its deals to other companies, it also began to cut corners on royalty payments, allegedly underreporting the volumes of gas it took from leases. More controversially, it shifted postproduction costs of natural gas to royalty owners. Such costs could include the expense of gathering, processing, and transporting gas to its buyer. Canny lessors included clauses in leases specifically barring the gas producer from passing these along, but many Barnett royalty owners had not had the foresight to insist on an exclusion from costs. No matter: in 2011, Chesapeake began deducting postproduction costs indiscriminately from royalty payments. Nor was that all. Royalty owners noted that Chesapeake actually sold the gas to one of its affiliates rather than to an independent purchaser, thus effectively allowing it to manipulate the price for which a royalty owner's gas was sold. With all these devices in play on Chesapeake's part, it is no wonder that some royalty owners saw their checks fall as much as 85 percent.[7]

Royalty owners responded by launching a veritable tsunami of litigation against Chesapeake. By the end of 2013, Chesapeake had been sued by landowners in Texas, Louisiana, Oklahoma, Pennsylvania, and Kansas; the DFW Airport; Fort Worth millionaire Ed Bass; the Cities of Fort Worth and Arlington; the Arlington Independent School District; and royalty owners in Pennsylvania. The US Department of the Interior's Office of Natural Resources Revenue levied fines for misreporting amounts of gas taken from Native American land. Chesapeake was under investigation relating to other issues by the US Securities and Exchange Commission, the Internal Revenue Service, and the US Department of Justice. Angry stockholders complained about a wide range of company maneuvers, including the size of McClendon's annual bonus, his sale of an expensive collection of maps to the company, and executive use of company jets for private purposes. They finally brought Chesapeake's board of directors to remove him as CEO in April 2013, but the company was not free from the courts, and neither was McClendon. When McClendon launched a new venture, American Energy Partners, he was sued by an Ohio-based coal company, American Energy Corporation's Century Mine, for deceptive trade practices and copyright infringement.[8]

For both Chesapeake and McClendon, far grimmer developments lay ahead. In an effort to regroup, Chesapeake directors replaced McClendon with Anadarko Petroleum executive Doug Lawler, who cut company spending by more than half and staffing by two-thirds. Cost-cutting notwithstanding, in 2015 the price of Chesapeake's shares fell 70 percent, the company wrote down $15.6 billion in assets, and it ended the year with a $4.15 billion net loss. It had already parted with over $30 billion in production and pipelines and put its Fort Worth office building up for sale. With respect to royalty litigation, Chesapeake fared no better. In 2012 it settled with DFW Airport for $5.3 million; in 2013 it settled with Pennsylvania royalty owners for $7.5 million, and in 2016 it settled with the City of Fort Worth for $6 million. The curtain fell on its spectacular involvement in the Barnett in August 2016, when it sold 215,000 acres of leases and 2,800 gas wells to French oil giant Total.[9]

Meanwhile, Aubrey McClendon plunged into a hectic campaign to build a second petroleum empire through his American Energy Partners. Never afraid of towering debt, he borrowed on a grand scale to pursue projected ventures from Oklahoma to Australia and Argentina. At least at first, his promotional magic worked as well as ever, and he found investors ready to direct dollars to his new creation. For all his boundless ambition, however, once oil prices headed downward in 2014, even his dauntless promotional skill was no match for discouraging industry economics. Despite a personal loan of over $400 million from banks, including Goldman Sachs Group, and his resorting to pledging personal assets—houses in Connecticut and Colorado, antique boats, his 20 percent stake in the NBA team the Oklahoma City Thunder, and two thousand bottles of vintage wines—as loan collateral, the money he desperately needed did not appear. Worse yet, unpleasant legal problems arose. In 2015 Chesapeake sued American Energy Partners, alleging McClendon had stolen proprietary maps and data. On March 1, 2016, a federal grand jury indicted him on charges of rigging oil and gas lease prices in Oklahoma, a charge he vehemently denied. On the morning of March 2, his car veered into a bridge abutment, killing him.[10] Thus ended the career of the most spectacular spender in the Barnett Shale.

While the Barnett Shale action died down in North Texas, the technology responsible for it transplanted to other areas, opening enormous new production of oil and gas. When gas drillers tweaked Barnett technology to tap the Marcellus Shale, which stretched over Pennsylvania and parts of West Virginia, New York, and Maryland, they began to produce huge reserves of natural gas, enough to meet the needs of millions of Middle Atlantic and New England consumers. The Marcellus easily eclipsed the Barnett as a source of natural gas. Underlying the Marcellus was another shale, the Utica, which produced both oil and gas in Ohio. Far more spectacular in terms of oil was the opening up of the Williston Basin's Bakken/Three Forks Shale, a prolific source of light sweet crude in western North Dakota, parts of South Dakota and Montana, and extending into Canada. North Dakota catapulted into the top ranks of US oil-producing states, going from production of roughly one hundred thousand barrels per day to over one million barrels per day in 2014. Millions of cubic feet of natural gas also flowed out of North Dakota Bakken wells, but lack of adequate pipeline outlets meant that much of it was flared. For that matter, lacking adequate pipelines meant that much of Bakken crude traveled to refineries by rail.[11] In both the Marcellus and Bakken plays, infrastructure could not handle the sudden and gargantuan leap in production. No one foresaw how prolific these shale plays could be.

The same problem of inadequate pipelines emerged when Barnett technology sparked dramatic new action in other parts of Texas, most notably in the South Texas Eagle Ford Shale and in the West Texas Permian Basin. Extending over twenty-three counties and potentially far beyond (the Eagle Ford formation is the source rock for the great East Texas oil field), Eagle Ford action opened in 2008 when Petrohawk Energy began to bring in gas wells averaging 8.9 million cubic feet per day (MMcfd) for the first day of production. Prospectors found that there were three producing areas in the Eagle Ford: natural gas production in its south, condensate in its middle, and oil in its north. Daily oil production went from several hundred barrels per day before 2008 to more than 1.6 million barrels per day at the beginning of 2015. At the end of 2014, its cumulative production reached one billion barrels from more than ten thousand completed wells. Eagle Ford wells also produced large quantities of natural gas, but an enormous amount was simply flared or vented—some 21 billion cubic feet (Bcf) according to one estimate.[12] Indeed, one of the major downsides of "oily shale" production has been the amount of gas wasted for lack of transportation to get it to market, resulting in part from continued low natural gas prices.

Permian Basin action took off when new fracking technology was modified and paired with horizontal drilling to escalate production from formations like the Spraberry, the Wolfcamp, and the Bone Springs. The Spraberry and Wolfcamp stretch over a large part of the Midland Basin in the eastern Permian Basin; the Bone Springs lies to the west in the Delaware Basin. Oilmen had retrieved millions of barrels of oil from formations above and below these, but wells either generated substantial production and then dwindled to insignificance shortly thereafter or were unpromising from the start. Thus operators usually drilled through the Wolfcamp to reach more lucrative targets; the Spraberry earned the title of the world's largest unrecoverable oil reserve. Drillers found that if they used massive, multiple fracks at successive levels in conventional vertical wellbores, they could bring in wells with hundreds of barrels of initial production. They got even larger yields with horizontal wellbores. The potential of the "Wolfberry" was stunning. In some places, for example, the Wolfcamp was one thousand feet thick—one thousand feet of oil-bearing rock stretching over some thirty counties! Production from this source alone quadrupled in a decade, reaching over 220,000 barrels per day in 2013. By March 2015, thanks in large part to production from rock previously too tight to be appealing, the Permian Basin yielded two million barrels per day.[13]

A great surge in production from shale meant that Texas's oil production dwarfed that of other regions. Indeed, in 2014, with production of oil at over 3.1 million barrels of oil per day, Texas was producing more oil than Saudi Arabia, the United Arab Emirates, and almost as much as Iraq. Were Texas an independent country, it would rank fourteenth in world crude production. As it was, Texas production helped national oil production increase to the highest levels seen in thirty-eight years, with a one-year growth in production in 2014 that was the highest since recordkeeping started in 1900.[14]

The surge in exploration and production generated by oilmen tapping shale in Texas and other booming shale regions meant skyrocketing growth in jobs, a growth that helped to offset the national recession in many areas. Of course, workers were essential in the field. In 2014, an estimated 302,700 workers in Texas had jobs in oil field drilling and service. There was a myriad of jobs producing what the oil field needed—electrical infrastructure, steel tubular goods, heavy equipment, chemicals, cement, and fracking sand. Trucks and truckers were in pressing demand; by the summer of 2013, the shortage of truck drivers in West Texas was so extreme that entry level wages for a trucker ran as high as $45,000 a year. Thousands of jobs also opened up in construction, hotels, restaurants, and retailing to serve the daily needs of all those workers prospering from shale-related employment. While it is not easy to calculate exactly how many people owed jobs, directly or indirectly, to the national impact of shale plays, the Texas

Independent Producers & Royalty Owners Association (TIPRO) estimated that in 2012, the oil and gas industry employed over 900,000 workers. The effect of action beyond the Barnett meant industry job growth of over 140,000 positions in the three years after 2009.[15] At the very least, in the gloomy national economic environment of 2008–14, shale-generated growth was a bright exception.

As was perennially true in the petroleum industry's history, however, oilmen's very success at producing tremendous amounts of oil and gas resulted in production beyond demand, which drove oil and gas prices downward. After 2008, natural gas prices failed to recover to levels stabilizing above $4/Mcf, not only acting as a damper on more gas drilling but also discouraging investment in infrastructure to salvage gas produced with oil from shale. And by mid-2014, the combination of bonanza oil production in the United States with faltering global oil demand caused oil prices to slide downward. Companies responded by cutting back projects and letting workers go; according to one estimate, by the end of July 2015, some 150,000 energy industry workers were out of work. Of course, low oil and gas prices brought benefits to consumers, from householders and vehicle owners, who paid less to heat homes and drive cars, to refiners, petrochemical and fertilizer manufacturers, and the airlines. The electric power industry had abundant incentive to proceed with switching from coal to natural gas. Consumption, however, did not rise enough to end oversupply.[16]

Enormous growth in national petroleum production raises the question of how long spectacular increase will continue. Just how large are the reserves and resources opened up by the new shale-focused technology? Will bumper production hold up? Is it even possible that national energy independence is in sight?

A variety of considerations make these questions hard to answer. One of the most important is the very novelty of shale production. The Barnett Shale in North Texas is as close to a mature shale-producing region as there is; it has been in production for less than two decades and is by no means all drilled up. Its outlying areas may become attractive exploration targets as technology develops. Areas like the Bakken and Eagle Ford are much newer. Their potential is still being defined. Moreover, like any oil or gas field, shale in a given area has sweet spots with impressive production and other places yielding less spectacular returns, making it more challenging to generalize about the ultimate recovery from a whole field.[17] One pattern that has emerged in shale production is that wells come in for strong initial production but rapidly decline in less than two years. Is the lower volume of production likely to stabilize at a fairly predictable rate or continue precipitous decline? If refracked, are old wells likely to return to yielding oil and gas at original levels? Here again, there may be substantial variations on a field-by-field or well-by-well basis. Over time, answers to some of these questions will become more obvious.

It is reasonable to assume, moreover, that over time the technology of bringing production from shale formations will steadily improve. In fact, operators drilling in the Permian Basin Wolfberry have already tweaked technology to drill wells faster, bring in bigger initial production at more sustainable rates, and lower costs—developments that have helped keep projects going at a time when oil prices have slumped. But the best technology cannot set aside economic reality. If oil and gas prices stay so low that there is little profit from exploration and production, or the technology is not affordable, the oil and gas in shale will stay underground as conventionally retrievable oil and gas did in the years leading up to this nation's energy crisis of the early 1970s. However great the reserves in shale, oilmen are not going to produce them at a loss. By mid-2015, production in the relatively high-cost Bakken and Eagle Ford Shale had already declined in response to the fall in crude prices.[18]

Notwithstanding the difficulty of assessing just how big reserves in shale are and how much petroleum production they will yield, the tantalizing vision of national energy independence has kept industry observers speculating about shale's energy potential. Skeptics have argued that current high levels of oil and gas production from shale are not going to last, certainly not for decades. Chief among them has been *World Oil* columnist Arthur Berman, who used a study of the Barnett

Shale he conducted in 2007–09 to argue against expecting shale gas production to hold up. Looking at some 2,000 Barnett wells, in 2009 he told a meeting of the Association for the Study of Peak Oil & Gas USA that ultimate gas recovery from the Barnett would only amount to 8.8 trillion cubic feet (Tcf), a far cry from USGS's estimate of 26 Tcf. Even wells in Barnett sweet spots declined dramatically in production during their first year or two, and reworking only led to "exponential terminal decline."[19] Nor had improved technology changed this discouraging picture. How could substantial Barnett production continue thirty or forty years, as optimists suggested, when by his estimate well production usually held up for less than eight years? Three years later he would remark that visions of shale production offering national energy independence were "preposterous."[20] Similarly, in 2013, author Bill Powers suggested that rather than offering a century of abundant energy, the nation would see a critical gas shortage within five to seven years. His book title, *Cold, Hungry and in the Dark: Exploding the Natural Gas Supply Myth* leaves no doubt of his point of view.[21] For advocates of replacing fossil fuels with renewable energy sources, the pessimists' message has been welcome.

But the shale pessimists are greatly outnumbered by the shale optimists, whose descriptive phrase of choice with respect to shale production has been "game changer." In 2010, researchers at IHS Cambridge Energy Research Associates (CERA) forecast that, thanks to shale, domestic natural gas supply would hold up for a century. They also estimated that the gas resource base (natural gas contained in all sources) would grow to over 3,000 Tcf. Daniel Yergin, CERA head and perhaps the best-known petroleum industry expert, shared these findings with *Wall Street Journal* readers the following year, predicting the United States would have enough natural gas to meet national demand for a century. In 2015 the Potential Gas Committee, a group composed of geoscientists and petroleum engineers, noted that between 2008 and 2014 the estimated size of the shale gas resource base more than doubled. When the size of shale gas resources was added to estimates for conventional gas and coalbed methane resources, the Potential

Gas Committee decided that the domestic natural gas resource base was 2,515 Tcf, an estimate compatible with those of IHS CERA researchers.[22]

In short, the optimists agree that the United States—for that matter, the world—has a tremendous amount of petroleum locked in shale. But just how tremendous is the amount in terms of resource base, and what kind of production is likely with existing technology? Estimates have ranged all over the statistical map, varying with times and methods of estimation. For example, in 2005 the US Energy Information Administration (EIA) put US recoverable shale gas reserves at 126 Tcf. In 2010, with Marcellus Shale exploration and production well under way, the EIA put its estimate of recovery just from the Marcellus at 410 Tcf, only to revise that estimate downward to 141 Tcf the following year. The EIA may have been responding to far more conservative estimates by the USGS. That agency hedged its estimates by saying ultimate Marcellus recovery would be anywhere from 43 to 144.1 Tcf of gas, giving its forecast wiggle room of over 100 Tcf! When the two agencies made their forecasts, however, they made strikingly different assumptions of how many wells would tap the Marcellus. The EIA assumed a drilling density of eight wells per square mile, the USGS of four wells per square mile.[23] Small wonder their estimates differed.

Not surprisingly, then, forecasts of the Barnett Shale's potential and ultimate natural gas recovery have varied considerably, even among those with relatively optimistic perspectives. Looking at a 5,000-square mile area, in 2003 the USGS estimated ultimate Barnett Shale gas recovery to be 26 Tcf. Working with a four-thousand-square mile area, the EIA put ultimate recovery at 23.81 Tcf in 2011. That same year, however, researchers at the Bureau of Economic Geology of the University of Texas at Austin undertook an exhaustive study of Barnett production. They looked at each of the 15,144 wells drilled between 1995 and 2010 and took into account variations in production from area to area, the impacts of changes in gas prices and drilling costs, and other recovery factors for each well. Their conclusions were more optimistic than the federal projections, looking to an ultimate recovery of

45 Tcf in the 4,172-square mile area that saw the most drilling. Broadening scope to an area of 8,000 square miles, they suggested that there might be 86 Tcf of recoverable gas, but only a small part of that area's gas had been developed. In terms of production longevity, they saw Barnett production peaking at 2 Tcf per year in 2012, but holding up through 2030, albeit declining to a production of only 900 Bcf per year. Even using a well-by-well approach, however, researchers admitted a wide variability among wells' ultimate production. What they considered a typical horizontal well might have a lifetime production ranging anywhere from 500 million cubic feet (Mcf) to over 4.3 Bcf.[24] Even a well-by-well study, then, left plenty of room for alternative recovery estimates.

If industry experts cannot agree on exactly how large shale petroleum resources are or how much oil and gas will be produced from wells drilled into shale formations, they do agree that, using current technology, oil and gas drillers are producing only a small fraction—perhaps less than one-tenth—of the petroleum locked in the rock. Depending on how refracking technology might be improved and deployed, estimates of ultimate recovery could rise dramatically. Out of some fifty thousand shale wells fracked since 2000, only six hundred have been refracked. It is too early to tell how successful attempts to revive old shale wells will be.[25]

So do oil and gas from shale offer the United States energy independence? Without a more definite idea of the extent of shale petroleum resources and how shale production holds up over time, this question is not easy to answer. The United States is producing enough natural gas to make imports of natural gas unnecessary. Indeed, the American gas industry would greatly benefit from exporting more of it. As for crude oil, imports reached a five-year low in 2014, bringing trade deficits from imported crude down 40 percent from what they were in 2010. Were American refiners able to engineer an overnight retrofit to refine light sweet shale crude as opposed to heavy sour foreign crude, the nation might go a long way toward short-term self-sufficiency. As it is, in 2015, American oil producers successfully pushed Washington to

repeal Nixon-era legislation prohibiting export of crude oil in order to take crude-swamping storage onto the global market.[26] Whether or not oil and gas from shale will add up to energy independence, they have decreased reliance on foreign petroleum. If optimists are right, they will continue to do so.

The prospect of this nation, which is the world's leading consumer of petroleum, being able to meet a significant portion of its gas and oil demand with domestic production has dramatic geopolitical implications in the global arena, as observers like Amy Myers Jaffe, Director for Energy and Sustainability at the University of California, Davis, have pointed out. Looking at natural gas, if the United States could meet domestic demand with its own gas, that would mean more gas on the global market for other consumers. Not only would that tend to keep global gas prices lower than otherwise, but if the United States could export a gas surplus, exports would keep global gas prices low. Cheaper energy in the form of abundant natural gas would help energy-hungry developing nations. At the same time, it would decrease the market power and geopolitical influence of such gas producers as Russia, Venezuela, and Iran, countries not known for pro-American foreign policy. An abundant global gas supply, moreover, would make it difficult to create a gas-producing cartel on the Organization of Petroleum Exporting Countries (OPEC) model, a cartel aimed at limiting supply to raise prices. Gas-consuming nations might meet their need for energy without buying from such a cartel—or they might be able to produce their own gas.[27]

As for the OPEC oil cartel, the surge in oil production in the United States has meant a striking erosion of OPEC's economic power in the global oil market, as well as the emergence of tensions within the cartel that threaten its continued existence. When OPEC was founded, its objective was to influence oil prices by limiting the supply of oil to nations like the United States that were dependent on what its members produced. Limiting supply in the face of rising demand would cause oil prices to rise, as they did in the 1960s and 1970s. But the weakness of this strategy lies in limiting supply. If consumer nations have alternative sources of oil, or if OPEC members cheat on their supply quotas, producing oil beyond agreed limits, OPEC's strategy does not work. By the same token, if an OPEC member nation like Saudi Arabia can produce an enormous amount of oil at very low cost, it may decide that it can accept low oil prices in return for enhanced market share, underselling higher-cost competitors. Such a move by a low-cost OPEC producer defeats OPEC strategy.

Since 2008, the huge rise in domestic oil production from shale has brought about a steady decrease in the amount of OPEC oil the United States has needed. US purchases of crude oil from OPEC members, for example, dropped over 14 percent in 2014 and another 16.5 percent in the first quarter of 2015. US imports of oil from Saudi Arabia, long America's second-largest supplier of foreign oil, fell 12.5 percent in 2014 and over 21 percent in the first quarter of 2015.[28] The oil the United States has not bought has swollen supply on the world market, helping prices slide to less than half what they were in mid-2014. In the face of these developments, OPEC is divided. Member nations like Venezuela, Libya, and Algeria, whose national budgets depend on revenue from oil sales, have desperately needed OPEC action that might limit world supply and raise prices. But even OPEC cutbacks have failed to raise prices as much as cartel members could hope, in part because of lagging global petroleum demand. [29]

The slide in global crude oil prices has indeed brought cutbacks and job losses in American shale oil-producing regions. Production is down in the relatively high-cost Bakken and Eagle Ford Shales. But oil operators have also responded to lower prices by enhancing efficiency. They have been more selective about where they drill; they have improved fracking technology to yield greater well production; and they have pressured service contractors to lower costs. As a result, in the shale oil fields in the Permian Basin, production rose during 2015–2016, as it has risen in many of the world's oil fields.[30] Thus Permian Basin producers' strategy in the face of lower prices has been simply to produce more oil.

In the contexts of the national economy and national security, the opening up of gas and oil production from shale has, overall, certainly been a game changer, and for the better. Opening up shale created thousands of jobs at a time of national recession, produced millions of dollars in revenue for local and state government, and lowered fuel and energy costs for consumers. It has decreased deficits in the national balance of payments and problems for both OPEC and other petroleum-producing countries at odds with our interests—Russia, for example. But if such benefits really have come at the cost of lasting damage to the environment, one can question if they are benefits at all. The opening of the Barnett Shale brought greater attention to a variety of environmental issues, especially those relating to air pollution from industry operations and fracking. What can we learn from the Barnett experience about them? Does producing gas and oil from shale mean inevitable environmental damage?

Any industry has at least some impact on the environment in which its operations take place, even if only changing the landscape. Those who bought homesteads in Wise or Denton Counties with a view to living a country life were understandably upset when drilling rigs appeared next door to them. They could not avoid the noise, lights, and fumes from equipment that go along with gas drilling. In general, gas drillers in the rural Barnett did little to make the industry presence in the countryside easier to accept. They failed to see that semisuburban communities and urban neighborhoods were not wide-open spaces where one could put rigs, processing units, or storage wherever one pleased. So country residents like Deborah Rogers found drilling rigs at work a few feet from their property lines. The residents of the hamlet of Dish suddenly had a major gas-processing hub barely outside city limits. Don Young learned that gas drillers planned to drill in the nature preserve across the street from his home. Barnett Shale development indicates that industry players often gave little thought to their neighbors.

Nor did the industry appear to take residents' concerns seriously. When Chesapeake wanted to run a gas pipeline through Carter Avenue front yards, the company was ready to ride roughshod over neighborhood objections until it met with a barrage of negative publicity. Sometimes companies ignored complaints or denied problems existed, as Deborah Rogers discovered. Similarly, when Dish homeowners invited gas company representatives to answer questions at a town meeting after a gas compressor malfunctioned and spewed natural gas into town, representatives from only two companies showed up and did nothing about complaints of noise and emission leaks. Looking at what the industry did not do, it is no wonder aggrieved Barnett Shale residents looked to regulatory agencies for help.

When residents turned to the Railroad Commission of Texas and the Texas Commission on Environmental Quality, however, they got little to no help. Perhaps one could explain the Railroad Commission's failure to handle Barnett Shale complaints by arguing that with thousands of drilling operations to monitor over the many counties in the Barnett Shale, the agency was simply out of its depth. It could not be everywhere at once; its staff and resources were inadequate to meet its regulatory obligations. Critics pointed out that in 2009, out of over 80,000 identified regulatory violations, the Railroad Commission levied only 379 penalties. If RRC shortcomings are understandable, if not excusable, it is harder to make a case for the TCEQ's handling of complaints. TCEQ was slow to act on complaints of air pollution from both Deborah Rogers and from Dish residents. When it did, it responded to residents' charges that emissions gave them health problems by suggesting there seemed to be nothing to worry about—yet. When the City of Fort Worth asked the TCEQ to investigate whether urban drilling was releasing carcinogens, especially benzene, into city air, the agency assured the city council that benzene was not a problem. Several months later the press revealed that the TCEQ had not used equipment sensitive enough to monitor the presence of benzene. At that point the city, like Rogers and the town of Dish, hired its own air analysts.

Were air-polluting emissions from industry operations making those living near them sick? The City of Dish decided to ask the Texas Department of State Health Services (DSHS) to investigate.

In December 2009, its staffers took blood and urine samples from twenty-eight adult residents to see if there were higher levels of volatile organic compounds in their blood and urine than 95 percent of the US population. They also sampled home tap water. Based on that small sample, the agency concluded there had not been "exposure to airborne contaminants, such as those that might be associated with natural gas drilling operations." Those whose blood and urine showed relatively high levels of harmful substances could have had sources of exposure like smoking or use of household cleaners, metal cleaners, or mothballs. Urine samples from Dish nonsmokers were similar to those of DSHS nonsmoking staffers in Austin. Only one home had contaminated tap water. The agency, then, did not find the community to be in danger. Still, it admitted there were many limitations in its investigation, one of them being that it could not identify with certainty the source for all exposures; it did not say why some people were sick.[31] Disturbingly, similar complaints of nosebleeds, headaches, and nausea surfaced in other locations, leaving open the possibility that emissions were a hazard to some individuals.

With good reason, Barnett Shale residents with concerns about industry operations found state agency responses unsatisfying and began to look to Washington for help from the EPA. At that point Barnett Shale problems entered national environmental controversy. But while attentive and, at least initially sympathetic, Washington did little that actually helped those with grievances. Rather the reverse. By appointing Al Armendariz as EPA regional administrator, Washington put forward a regulator whose obvious anti-industry bias would discredit both the EPA and the regional environmental cause. Here, as we have seen, the issue of fracking would challenge the credibility of both Armendariz and the EPA.

Does the Barnett Shale experience support the claim that fracking causes environmental damage? Thus far, despite a multitude of investigations on the part of regulatory agencies and industry consultants, no one has found a specific instance in which fracking rock thousands of feet below a water source has contaminated that water source.

Apart from natural oil and gas seeps that are common in many regions, including North Texas, there are many ways in which petroleum can foul ground water. It can move into water sources from abandoned, improperly plugged wells; it can leak from a producing well as a result of casing or cementing failure. The likeliest cause of water contamination connected with fracking, however, is irresponsible disposal of the fluid used to frack a well. Rather than truck frack water to disposal wells, some irresponsible operators have dumped fluid returning from fracked wellbores into streams, creeks, or municipal sewage systems. In some instances, disposal truckers have simply released frack water as they drove down highways, allowing it to contaminate soil and the groundwater beneath it. Subsequent water pollution in those cases were the result of what happened above the ground rather than below it.

Disposal wells in the Barnett Shale and other areas are the suspected source of another, quite unexpected problem—earthquakes. Although it has been difficult to prove, it is possible that injecting millions of gallons of frack water down a disposal well located near an underground rock fault may cause shifts in the rock. Some investigators have argued that disposal wells at DFW Airport and near the Parker County communities of Reno and Azle have caused earth tremors. An increased instance of earthquakes in Oklahoma has led the Oklahoma Corporation Commission to limit amounts of frack water injected down disposal wells.

Indeed, recent evidence indicates that disposal wells should be located with more care than has hitherto been usual. According to a study published by *Science Advances* in November of 2017, "in the Fort Worth Basin, along faults that are currently seismically active, there is no evidence of prior motion over the past (approximately) 300 million years." The study was conducted by a team led by Beatrice Magnani, associate professor of geophysics in SMU's Huffington Department of Earth Sciences. It was coauthored by Michael L. Blanpied, associate coordinator of the US Geological Survey's Earthquake Hazard program, and SMU seismologists Heather DeShon and

Matthew Hornbach. "The study's findings suggest that the recent Fort Worth Basin earthquakes, which involve swarms of activity on several faults in the region, have been induced by human activity," according to Blanpied.

Disposal problems, of course, can be avoided if water used in fracking is recycled, used to frack many wells rather than one. The technology to remove mineral salts and organic compounds from frack water has emerged in the course of Barnett Shale drilling. Used successfully by operators like Devon Energy, it permits recycling most of the water used and thus addresses concerns about depleting water supply in drought-prone regions. Barnett development has also seen improvements in technology to reduce waste of natural gas in well completion, so-called "green completions," which can decrease air pollution from natural gas emissions. In the Barnett context, then, advances in technology have offered the prospect of greater protection against potential environmental damage.

Of course, if one opposes continued use of fossil fuels like oil and gas, seeing it as inherently damaging to the environment, one will have a negative view of both the opening up of the Barnett Shale and the technology making it possible. Without a doubt, the availability of an abundant supply of gas and oil from shale has worked against a sense that it is urgent to develop and shift to renewable sources of energy. But another way of approaching petroleum production from shale is to see it as a way, over the long term, to ease the transition to use of renewables—a relatively inexpensive way to meet national energy demand until renewable sources of energy are cheaper than at present.

Directly or indirectly, millions of Americans have benefitted from what the Barnett Shale boom started. The boom itself created thousands of jobs for workers both inside and outside the petroleum industry. By resulting in a superabundance of natural gas, it lowered the cost of raw material for some industries and of fuel for many others, encouraging manufacturing growth and creation of yet more jobs. But perhaps the greatest beneficiaries of shale development are the Americans who live on limited incomes. For the working poor, the elderly on fixed incomes, the unemployed, and the disabled, paying less to fill up a car, heat a home, or keep lights on really matters. In this respect, cheaper energy from shale has helped those who have needed the most help. They should be kept in mind when looking at what the Barnett Shale boom began.

NOTES

INTRODUCTION

1. Kenneth S. Deffeyes, *Hubbert's Peak: The Impending World Oil Shortage* (Princeton: Princeton University Press, 2009), 1–4.
2. Roger M. Olien and Diana Davids Olien, *Oil and Ideology: The Cultural Creation of the American Petroleum Industry* (Chapel Hill: University of North Carolina Press, 2000), 121–125, 192–193.
3. Deffeyes, *Hubbert's Peak*, xi. Similarly, in 2005, Deffeyes's year of peak global production, Matthew R. Simmons argued that pessimism about global oil supply was justifiable because it was unlikely that Saudi Arabia could continue to product oil at high rates as its oil fields aged; Matthew R. Simmons, *Twilight in the Desert: The Coming Saudi Oil Shock and the World Economy* (New York: John Wiley & Sons, Inc., 2005), xvi–xvii.
4. "U.S. Oil and Gas Drilling Boom's Economic Impact," *Midland Reporter-Telegram*, May 21, 2015; Jeffrey Sparshott, "Oil Boom a 'Game-Changer' on Trade Deficit," *Wall Street Journal*, February 6, 2015; Jennifer Hiller, "American Crude Hits a 114–year High, Led by Shale," *Midland Reporter-Telegram*, April 6, 2015; Robert Grattan, "Industry Passes the All-Time High for Inventories," *Midland Reporter-Telegram*, May 27, 2015.
5. Jim Malewitz, "Study: Barnett Shale Richer than Previously Thought," *Midland Reporter-Telegram*, December 20, 2015. Estimates of the size of Barnett reserves, however, have been highly variable. In 2013, a group of geoscientists offered a figure of 45 trillion cubic feet; John Browning,

Scott W. Tinker, Svetlana Ikonnikova, Gurcan Gulen, Eric Potter, Qilong Fu, Susan Horvath, Tad Patzek, Frank Male, William Fisher, Forrest Roberts, and Ken Medlock III, "Study Develops Decline Analysis, Geologic Parameters for Reserves, Production Forecast," *Oil & Gas Journal* (2013): 62. But a year later, Rafael Sandrea and Ivan Sandrea suggested the Barnett's ultimate recovery would be twenty trillion cubic feet, with a gas in place estimate of an average of fifty trillion cubic feet per square mile; "New Well-Productivity Data Provide US Shale Potential Insights," *Oil & Gas Journal* (2014).

CHAPTER 1

1. For an excellent recent study of the Cross Timbers region, see Richard V. Francaviglia, *The Cast Iron Forest: A Natural and Cultural History of the North American Cross Timbers* (Austin: University of Texas Press, 2000), 1–13, 76–81, *et passim*. See also Terry G. Jordan, with John L. Bean, Jr., and William M. Holmes, *Texas: A Geography* (Boulder: Westview Press, 1984), 31, 35–36; Washington Irving, *A Town on the Prairies*, edited and with an Introductory Essay by John Francis McDermott (Norman: University of Oklahoma Press, 1956), 125.
2. Francaviglia, *The Cast Iron Forest*, 112–116; Jordan, *Texas: A Geography*, 16, 34. A useful introduction to the histories of the counties in this region may be found as entries in Ron Tyler et al., eds., *The New Handbook of Texas* (Austin: The Texas State Historical Association, 1996). See, for example, "Clay County," vol. 2, 146; "Denton County," vol. 2, 599; "Jack County," vol. 3, 890; Johnson County," vol. 3, 966–967; "Palo Pinto County," vol. 5, 29–30; "Parker County," vol. 5, 63–64; "Wise County," vol. 6: 1028–1029; Randolph B. Campbell, *Gone to Texas: A History of the Lone Star State* (New York: Oxford University Press, 2003), 216.
3. Stephen Powers, *Afoot and Alone: A Walk from Sea to Sea by the Southern Route, Adventures and Observations in Southern California, New Mexico, Arizona, Texas, Etc.*, edited and with an introduction by Harwood P. Hinton (Austin: The Book Club of Texas, 1995), 115–116.
4. Francaviglia, *The Cast Iron Forest*, 144, 163–165; John S. Spratt, *The Road to Spindletop: Economic Change in Texas, 1875–1901* (Austin: University of

Texas Press, 1970), 81–82; Tyler et al., *New Handbook of Texas*, "Boll Weevil," vol. 1: 628–629.

5. United States. Department of the Interior. United States Geological Survey, *Water Supply Paper 317*: C. H. Gordon, "Geology and Underground Waters of the Wichita Region, North-Central Texas" (Washington, DC: Government Printing Office, 1913), 34, 37, 73, 78; Spratt, *Road to Spindletop*, 261–264. Much has been written about Thurber. Three especially helpful works are Don Woodard, *Black Diamonds! Black Gold!* (Lubbock: Texas Tech University Press, 1998); John S. Spratt Sr., *Thurber, Texas: The Life and Death of a Company Coal Town*, ed. Harwood P. Hinton (Abilene: State House Press, 2005); and Marilyn D. Rhinehart, *A Way of Work and a Way of Life: Coal Mining in Thurber, Texas, 1888–1926* (College Station: Texas A&M University Press, 1992).
6. Gordon, *Geology and Underground Waters*, 11, 34; Catherine Troxell Gonzalez, *Rhome: A Pioneer History* (Burnet: Eakin Press, 1979), 31; *Decatur News*, July 10, 1908.
7. University of Texas. Bureau of Economic Geology. University of Texas Bulletin 2544: W. M. Winton, *The Geology of Denton County* (Austin: University of Texas, 1925), 14–16; *Decatur News*, September 20, 1907; *Fort Worth Star-Telegram*, June 27, 1923; United States. Department of Interior. United States Geological Survey, Bulletin 629: *Natural Gas Resources of Parts of North Texas*, Eugene Wesley Shaw, "Gas in the Area North and West of Fort Worth" (Washington, DC: Government Printing Office, 1916), 54.
8. Frank J. Gardner, *Rinehart's North Texas Oil: A Correlation of Characteristics of the Oil Fields of North and North Central Texas* (Houston: Rinehart Oil News Company of Texas, 1941), 17; *Oil & Gas Journal*, April 28, 1920, 64. For a lengthy survey of oil and gas shows in North Texas, see C. A. Warner, *Texas Oil & Gas Since 1543* (Houston: Gulf Publishing Company, 1939), 226–227.
9. Warner, *Texas Oil*, 226–227; George H. Fancher with Robert L. Whiting and James H. Cretsinger, *The Oil Resources of Texas: A Reconnaissance Survey of Primary and Secondary Reserves of Oil* (Austin: The Texas Petroleum Research Committee, 1954), 333; Frank J. Gardner, *Rinehart's North Texas Oil*, 17.
10. Diana Davids Olien and Roger M. Olien, *Oil in Texas: The Gusher Age, 1895–1945* (Austin: University of Texas Press, 2002), 4–5, 9; For the fullest

account of Cullinan and Corsicana, see John O. King, *Joseph Stephen Cullinan: A Study of Leadership in the Texas Petroleum Industry, 1897–1937* (Nashville: Vanderbilt University Press, 1970), 29–46.

11. Olien and Olien, *Oil in Texas*, 24–35; James A. Clark and Michel T. Halbouty, *Spindletop* (New York: Random House, 1952), 38–42, 128–132. For Pattillo Higgins's career see Robert W. McDaniel with Henry C. Dethloff, *Pattillo Higgins and the Search for Texas Oil* (College Station: Texas A&M University Press, 1989). The story of Spindletop can be found in Judith Walker Linsley, Ellen Walker Rienstra, and Jo Ann Stiles, *Giant Under the Hill: A History of the Spindletop Oil Discovery at Beaumont, Texas, in 1901* (Austin: Texas State Historical Association, 2002).
12. McDaniel and Dethloff, *Pattillo Higgins*, 47; Edgar Wesley Owen, *Trek of the Oil Finders: A History of Exploration for Petroleum* (Tulsa: American Association of Petroleum Geologists, 1975), 195; Clark and Halbouty, *Spindletop*, 142; Olien and Olien, *Oil in Texas*, 27, 43.
13. Olien and Olien, *Oil in Texas*, 76–77; Gordon, *Geology and Underground Waters*, 45–46; Francher, *Oil Resources*, 325; Warner, *Texas Oil*, 49.
14. Olien and Olien, *Oil in Texas*, 77; Warner, *Texas Oil*, 51–52, 227–228; Gardner, *Rinehart's North Texas Oil*, 19. The companies, a part of this Corsicana-based group, would evolve into part of Magnolia Oil, which, in turn, would be part of the Mobil (Standard Oil of New York) organization. Because Standard Oil was under legal attack in Texas at this time, Payne and his friends were careful to present their involvement in Texas oil as that of individual investors.
15. The first American use of natural gas for illumination took place when residents of Fredonia, New York, used natural gas from a twenty-seven-foot well for lighting in 1821; James A Clark, *The Chronological History of the Petroleum and Natural Gas Industries* (Houston: Clark Book Co., 1963), 15.
16. Gardner, *Rinehart's North Texas Oil*, 19–21; Warner, *Texas Oil*, 53; Olien and Olien, *Oil in Texas*, 78.
17. Olien and Olien, *Oil in Texas*, 78–79; Gardner, *Rinehart's North Texas Oil*, 21; Warner, *Texas Oil*, 228–229; Ralph O. Harvey Jr., interviewed by Diana Davids Olien and Roger M. Olien, July 25, 1996, Wichita Falls, Texas.
18. Warner, *Texas Oil*, 229.
19. J. A. Udden and Drury McN. Phillips, *A Reconnaissance Report on the Geology of the Oil and Gas Fields of Wichita and Clay Counties, Texas*, Bureau of Economic Geology and Technology, University of Texas, Bulletin 246

(Austin: University of Texas, 1912), 103; Gardner, *North Texas Oil*, 62; Richard Mason, "Lost Seas and Forgotten Climes: Petroleum and Geologists in North Texas," *West Texas Historical Association Year Book* 63 (Lubbock: Texas Tech University Press, 1987), 135–138. Mason's article offers an excellent overview of geologists' approaches to North Texas.

20. Udden and Phillips, *Reconnaissance Report*, 62–63.
21. Gardner, *Rinehart's North Texas Oil*, 21.
22. Don Woodward, *Black Diamonds! Black Gold! The Saga of Texas Pacific Coal and Oil Company* (Lubbock: Texas Tech University Press, 1997), 100–107; F. W. Reeves and W. C. Bean, "The Ranger Oil and Gas Field," in L. C. Snider, *Oil and Gas in the Mid-Continent Fields* (Oklahoma City: Harlow Publishing Co., 1920), 261–265.
23. In *Oil Booms: Social Change in Five Texas Towns* (Lincoln: University of Nebraska Press, 1952), Roger M. Olien and I have suggested the Ranger boom's excesses were greatly exaggerated by journalist Boyce House and others; see *Oil Booms*, 10, 127–128, 141–142. See also Reeves and Bean, "Ranger Oil Field," 264; Olien and Olien, *Oil in Texas*, 81–83; Owen, *Trek of the Oil Finders*, 318–319.
24. Woodard, *Black Diamonds!*, 103; H. A. Wheeler, "Wild Boom in the North Texas Oil Fields," *Engineering and Mining Journal* (1920): 741–747.
25. Olien and Olien, *Oil in Texas*, 92–93. On conservationist criticism of Burkburnett, see Roger M. Olien and Diana Davids Olien, *Oil and Ideology: The Cultural Creation of the American Petroleum Industry* (Chapel Hill: University of North Carolina Press, 2000), 154–162. See also L. W. Kesler, "The Burkburnett Oil Field," in Snider, *Mid-Continent Fields*, 246–249.
26. *Wise County Messenger*, February 28, 1913; *Decatur News*, June 13, 1913; *Wise County Messenger*, April 3, June 18, 1920; *Oil Weekly*, February 8, 1919, 36, March 1, 1919, 40; *Fort Worth Record*, February 16, 1919; *Decatur News*, June 11, 1920; *Wise County Messenger*, June 4, 1920; *Fort Worth Star-Telegram*, October 31, 1922.
27. *Fort Worth Record*, January 21, 1920; *Fort Worth Star-Telegram*, November 2, 1922, July 8, August 15, 1923.
28. Janet L. Schmelzer, *Where the West Begins: Fort Worth and Tarrant County* (Northridge: Windsor Publications, Inc., 1985), 62; J'Nell Pate, *Livestock Legacy: The Fort Worth Stockyards, 1887–1987* (College Station: Texas A&M University Press, 1988), 10, 13–15, 121; Cissy Stewart Lale, in Oliver Knight, *Fort Worth: Outpost on the Trinity* (Fort Worth: TCU Press, 1990), 114–115,

193; The Historical Committee of the Fort Worth Petroleum Club, *Oil Legends of Fort Worth* (np: Taylor Publishing Company, 1993), 41, 74–77; Robert H. Talbert, *Cowtown-Metropolis: Case Study of a City's Growth and Structure* (Fort Worth: TCU Press, 1956), 36–37.

29. *Oil Legends*, 37–40; Olien and Olien, *Easy Money: Oil Promoters and Investors in the Jazz Age* (Chapel Hill: University of North Carolina Press, 1990), 75; Talbert, *Cowtown-Metropolis*, 36–39.
30. For brief biographies of these oilmen, see *Oil Legends*, 102–103, 151, 154–155, 119–120, 191–192, 176–177, 103–104, 208, 164–166, 213–215, and 21–23 for the Golden Goddess. On Ed Landreth, see Roger M. Olien and Diana Davids Hinton, *Wildcatters: Texas Independent Oilmen* (College Station: Texas A&M University Press, 2002), 28–35.
31. Olien and Olien, *Easy Money*, 76–81, 92–95, 104–107; Wheeler, "Wild Boom," 742.
32. Olien and Olien,147–167.

CHAPTER 2

1. As a technique of well stimulation, hydraulic fracturing was first used in the Hugoton field in southwestern Kansas in 1947; Hilton Price, "Frac, Rinse, Repeat," PennEnergy, http://www.pennenergy.com/index/articles/display/5563829146/articles/pennenergy, accessed April 8, 2012.
2. On postwar natural gas industry growth, see Christopher J. Castaneda and Clarence M. Smith, *Gas Pipelines and the Emergence of America's Regulatory State: A History of Panhandle Eastern Corporation, 1928–1993* (New York: Cambridge University Press, 1996), 4–8, 124–128; Christopher J. Castaneda and Joseph A. Pratt, *From Texas to the East: A Strategic History of Texas Eastern Corporation* (College Station: Texas A&M University Press, 1993), 33–81.
3. John H. Stovall, Leslie W. Dorbandt, and Walter L. Ammon, "Developments in North and West-Central Texas in 1947," *Bulletin of the American Association of Petroleum Geologists* (hereafter cited as *AAPG Bulletin*) 32, no. 6 (June 1947), 988–996; Walter L. Ammon, L. W. Dorbandt, and John H. Stovall, "Developments in North and West-Central Texas in 1948," *AAPG Bulletin* 33:6 (June 1949), 935–937; H. H. Bradfield, "Developments in North Texas in 1950," *AAPG Bulletin* 35, no. 6 (June 1951), 1289–1290; *Oil*

& *Gas Journal*, December 27, 1947, 297, September 16, 1948, 161, January 4, 1951, 96. See also K. S. Blanchard, Orval Denman, and A. S. Knight, "Natural Gas in Atokan (Bend) Section of Northern Fort Worth Basin," in B. Warren Beebe, ed., *Natural Gases of North America: A Symposium in Two Volumes*, volume 2 (Tulsa: American Association of Petroleum Geologists, 1968), 1446–1452; and George Glover, "A Study of the Bend Conglomerate in S. E. Maryetta Area, Boonsville Field, Jack County, Texas: in Charles A. Martin, ed., *Petroleum Geology of the Fort Worth Basin and Bend Arch Area* (Dallas: Dallas Geological Society, 1982), 353–354.

4. Joseph W. Kutchin, *How Mitchell Energy and Development Corp. Got Its Start and How It Grew: An Oral History and Narrative Overview* (The Woodlands: Mitchell Energy & Development Corp., 1998), 183–186.
5. Kutchin, 186–188.
6. Kutchin, 191–192, 225–228.
7. George P. Mitchell, interviewed by Diana Davids Hinton, July 7, 2010, Houston, Texas; Kutchin, *Mitchell Energy*, 2–6; Dan B. Steward, *The Barnett Shale Play: Phoenix of Fort Worth Basin: A History* (Fort Worth: Fort Worth Geological Society, North Texas Geological Society, 2007), 26.
8. Mitchell interview; Kutchin, *Mitchell Energy*, 6, 193; Steward, *Barnett Shale Play*, 26.
9. Kutchin, *Mitchell Energy*, 229.
10. Mitchell interview; Kutchin, *Mitchell Energy*, 229; Steward, *Barnett Shale Play*, 26–28.
11. Kutchin, *Mitchell Energy*, 63–67; Steward, *Barnett Shale Play*, 27.
12. On the problems independents faced and how they met them, see Roger M. Olien and Diana Davids Hinton, *Wildcatters: Texas Independent Oilmen* (College Station: Texas A&M University Press, 2007), chapter 6. Intrastate gas prices were free of federal regulation at this time, and they rose higher than interstate prices. However, once a gas producer like Mitchell directed gas to the interstate market, federal regulation prohibited redirecting that gas to sale in the intrastate market. On federal natural gas regulation, see Richard H. K. Vietor, *Energy Policy in America since 1945: A Study of Business-Government Relations* (Cambridge: Cambridge University Press, 1984), chapters 4 and 7. For a comprehensive history of the US natural gas industry through 1995, see Arlon R. Tussing and Bob Tippee, *The Natural Gas Industry: Evolution, Structure, and Economics*, second edition (Tulsa: Pennwell Books, 1995).

13. Kutchin, *Mitchell Energy*, 6–8, 70, 198.
14. For a succinct introduction to the 1970s energy crisis, see Karen R. Merrill, *The Oil Crisis of 1973–1974: A Brief History with Documents* (Boston: Bedford/St. Martin's, 2007).
15. John C. McCaslin, "Fort Worth Basin Activity Hums," *Oil & Gas Journal*, September 16, 1974, 137; Jim West, "High Gas Prices Trigger N. Texas Lease Play," *Oil & Gas Journal*, November 5, 1975, 105.
16. Renee Wash, "Area Holding its Own," *Drill Bit*, April 1983, 27.
17. Jerry Hodgden and C. A. Martin, "Fort Worth Basin Very Busy Again," *Oil & Gas Journal*, November 11, 1974, 250.
18. Dick Lowe, interviewed by Diana Davids Hinton, July 27, 2010, Fort Worth, Texas; Ted Collins, interviewed by Diana Davids Hinton, July 23, 2009, Midland, Texas.
19. Steward, *Barnett Shale Play*, 30–32; "Mitchell Begins Drilling Project in Texas Lake," *Oil & Gas Journal*, July 6, 1981, 74; Vernetta Mickey, "Mitchell EOR Projects Yield Tertiary Oil in Wise and Jack Counties," *Drill Bit*, September 1982, 45; Stephen A. Holditch and David E. Lancaster, "Economics of Austin Chalk Production," *Oil & Gas Journal*, August 9, 1982, 183.
20. Tussing and Tippee, *Natural Gas Industry*, 193–195; Kutchin, *Mitchell Energy*, 9, 31; Steward, *Barnett Shale Play*, 31; Dan B. Steward, interviewed by Diana Davids Hinton, March 4, 2010, Dallas, Texas. George Mitchell's opportunistic use of federal energy measures would support Paul Sabin's argument that political realities shape economic outcomes; see his *Crude Politics: The California Oil Market, 1900–1940* (Berkeley: University of California Press, 2005). For that matter, as president of the Texas Independent Producers and Royalty Owners Association (TIPRO) in the early 1970s, Mitchell himself did what he could to shape federal energy policy; see Lawrence Goodwyn, *Texas Oil, American Dreams: A Study of the Texas Independent Producers and Royalty Owners Association* (Austin: Texas State Historical Association, 1996), 121–122, 223. The federal tax credit on tight sand gas continued until 1993, at which time it was continued by the state of Texas as a credit against severance tax.
21. Kent A. Bowker, "Development of the Barnett Shale, Fort Worth Basin," *Search and Discovery* Article #10126, posted April 18, 2007, http://www.searchanddiscovery.net/2007/07023/bowker/index.htm; Joseph H. Frantz Jr., George A. Waters, and Valerie Jochen, "Operators Rediscover Shale Gas Value," *E & P*, October 1, 2005, http://www.epmag.com/archives/print/3846.htm.

22. Steward, *Barnett Shale Play*, 32; James D. Henry, "Stratigraphy of the Barnett Shale (Mississippian) and Associated Reefs in the Northern Fort Worth Basin," in Charles A. Martin, ed. *Petroleum Geology of the Fort Worth Basin and Bend Arch Area* (Dallas: Dallas Geological Society, 1982), 157.
23. Steward interview; Steward, *Barnett Shale Play*, 37–38, 45.
24. Steward, *Barnett Shale Play*, 38, 44, 50–51; Steward interview; Dan B. Steward, "Evolution of the Barnett Shale Play," *E & P*, March 19, 2008, http://www.epmag.com/article/print/3653.
25. Steward, *Barnett Shale Play*, 48–49, 60–61; Steward interview; Vernetta Mickey, "Austin Chalk Production Depends on Connecting Natural Fracs to Wellbore," *Drill Bit*, May 1982, 72–74.
26. Steward, *Barnett Shale Play*, 74. On the gas bubble, see Tussing and Tippee, *Natural Gas Industry*, 196–199.
27. Steward interview; Steward, *Barnett Shale Play*, 66–67, 92, 96.
28. Steward interview.
29. Steward, *Barnett Shale Play*, 110–113.
30. Steward interview; Larry Brogdon, interviewed by Diana Davids Hinton, July 28, 2010, Fort Worth, Texas; Steward, *Barnett Shale Play*, 125.
31. Steward, *Barnett Shale Play*, 127–129.
32. Steward, *Barnett Shale Play*, 129, 134; Steward interview.
33. Steward, *Barnett Shale Play*, 92–94, 129–132, 146–149; Jim Fuquay, "Q & A George Mitchell," *Fort Worth Star-Telegram*, April 2, 2008.
34. Steward, *Barnett Shale Play*, 165–169.
35. Steward interview.
36. Ken Morgan, interviewed by Diana Davids Hinton, July 30, 2010, Fort Worth, Texas.
37. Brogdon interview.
38. Charles Moncrief, interviewed by Diana Davids Hinton, July 27, 2009, Fort Worth, Texas.
39. Brogdon interview.

CHAPTER 3

1. Steward, *Barnett Shale Play*, 37, 175; Jack Z. Smith, "Tarrant Sites Attract Plans for Drilling," *Fort Worth Star-Telegram*, April 12, 1993; Jack Z. Smith, "Two Large Firms Gauge the Merit of Natural Gas Wells in the County's Far Northwest Section," *Fort Worth Star-Telegram*, June 24, 1993.

2. Steward interview; Steward, *Barnett Shale Play*, 176–177; Kent A. Bowker, "Barnett Shale Gas Production, Fort Worth Basin: Issues and Discussion," *AAPG Bulletin* (April 2007), 223–225.
3. Brogdon interview; Steward, *Barnett Shale Play*, 85, 174.
4. Vello A. Kuuskraa, George Koperna, James W. Schmoker, and John C, Quinn, "Barnett Shale Rising Star in Fort Worth Basin," *Oil & Gas Journal*, May 25, 1998.
5. Brogdon interview; Kuuskraa, "Barnett Shale Rising Star," 67–71; Steward, *Barnett Shale Play*, 175.
6. Kathy Shirley, "Barnett Shale Living Up to Potential," *AAPG Explorer*, July 2002, 27; Brogdon interview.
7. Brogdon interview.
8. Dick Lowe, interviewed by Diana Davids Hinton, July 27, 2010, Fort Worth, Texas; Brogdon interview.
9. Brogdon interview; Steward, *Barnett Shale Play*, 178–179.
10. Brogdon interview.
11. Brogdon interview; Lowe interview; Marty Searcy, interviewed by Diana Davids Hinton, February 20, 2009, Fort Worth, Texas.
12. Searcy interview; Brogdon interview; Lowe interview; Robert Francis, "Four Sevens Oil Rolls Barnett Shale Dice Again," *Fort Worth Business Press*, June 12, 2006.
13. Lowe interview; Brogdon interview; Searcy interview; Dan Piller, "Pair Seek Analysis of Valuation," *Fort Worth Star-Telegram*, March 21, 2006; Robert Francis, "Four Sevens Oil rolls Barnett Shale Dice Again," *Fort Worth Business Press*, June 12, 2006.
14. Leslie Haines, "Barnett Marvel," *Oil and Gas Investor*, January 2007, 74; George M. Young Jr., interviewed by Diana Davids Hinton, July 28, 2009, Fort Worth, Texas.
15. Young interview; Ted Collins, interviewed by Diana Davids Hinton, July 23, 2009, Midland, Texas.
16. Young interview.
17. Young interview; Collins interview; Haines, "Barnett Marvel," 75; Peggy Williams, "The Barnett Shale," *Oil and Gas Investor*, 40–43; Michael Whitely, "Chief Oil, Hillwood Partners in Alliance Natural Gas Well Project," *Dallas Business Journal*, August 16, 2003.
18. Young interview; Collins interview; Dan Piller, "Deal Highlights Growing Wealth in Barnett Shale Play," *Fort Worth Star-Telegram*, May 3, 2006;

"Chief Sold," *Fort Worth Business Press*, May 8, 2006; Holli L. Estridge, "CrossTex Says Chief Assets Fit Into Growth Plan," *Dallas Business Journal*, June 30, 2006.

19. Hollis Sullivan, interviewed by Diana Davids Hinton, July 27, 2010, Fort Worth, Texas.
20. Sullivan interview.
21. Sullivan interview.
22. Sullivan interview.
23. Sullivan interview.
24. Sullivan interview; Steward, *Barnett Shale Play*, 180; Robert Francis, "Pump It Up," *Fort Worth Business Press*, July 23, 2004; Robert Francis, "Barnett Shale Makes Deep Impact on Oil, Gas Markets," *Fort Worth Business Press*, March 28, 2005.
25. "Barnett Shale Area Expansion Expected in 2005," *Oil & Gas Journal*, March 7, 2005, 43; Young interview.
26. John-Laurent Tronche, "Exxon Mobil's Historic Buy," *Fort Worth Business Press*, December 21, 2009; "Barnett Shale Area Expansion," 42; Dan Piller, "XTO Buys More Natural Gas Leases," *Fort Worth Star-Telegram*, June 2, 2006; Robert Francis, "XTO Energy Makes Its Largest Acquisition to Date," *Fort Worth Business Press*, June 11, 2007; Ben Casselman, "XTO Strikes $4.19 Billion Hunt Deal," *Wall Street Journal*, June 11, 2008.
27. Sullivan interview; Searcy interview; "Barnett Shale Area Expansion," 43. For Chesapeake's plunge into leasing thousands of acres at top prices in the Barnett, see Russell Gold, *The Boom: How Fracking Ignited the American Energy Revolution and Changed the World* (New York: Simon & Schuster, 2014), 191–197.
28. Young interview; Lowe interview; Charles Moncrief interview.
29. Alan Petzet, "Devon Pressing Barnett Shale Exploitation, Expanding Search," *Oil & Gas Journal*, July 15, 2002, 18–20; "Devon, EOG Show Barnett Shale Has Momentum," *Oil & Gas Journal*, August 8, 2005, 25–26.
30. "Small Barnett Shale Producers Cash Out," *Fort Worth Star-Telegram*, June 6, 2006; Russell Gold, "Big Oil Firms Join Hunt for Natural Gas in U.S.," *Wall Street Journal*, November 29, 2005; "Pioneer Resources Buys Shell Oil's Barnett Shale Holdings," *Fort Worth Star-Telegram*, November 6, 2007; Charles Moncrief interview.
31. Elizabeth Souder, "Exxon Mobil Enters Barnett Shale Deal," *Denton Record Chronicle*, January 30, 2007; Robert Francis, "Exxon Mobil subsidiary Dips

Toe in Barnett Shale," *Fort Worth Business Press*, February 5, 2007; Charles Moncrief interview.

CHAPTER 4

1. Steward, interview; Holli Estridge, "Barnett Shale: Fuel for the Economy," *Dallas Business Journal*, June 8, 2007; "Barnett Play Moves Far South of Fort Worth," *Oil & Gas Journal*, June 11, 2007, 8.
2. L. Decker Dawson, C. Ray Tobias, and Steve C. Jumper, interview by Diana Davids Hinton, July 6, 2009, Midland, Texas. With respect to seismic work, Ray Tobias noted, "We're used to define what to stay away from." Steward, interview; Brogdon, interview; Steward, *Barnett Shale Play* 184–185; "EOG Makes Barnett Shale Oil Discovery," *Oil & Gas Journal*, March 3, 2008, 7.
3. Estridge, "Barnett Shale;" Ann Zimmerman, "A Lot of Gas: Wildcatter Sanford Dvorin Thinks He's Hit the Jackpot in Quaint Coppell," *Dallas Observer*, May 15, 1997; Eric Aasen, "Gas Drilling to Expand in Irving," *Dallas Morning News*, February 9, 2007; Jim Fuquay, "Barnett Shale Drillers Going Farther Afield," *Fort Worth Star-Telegram*, July 27, 2008.
4. Bernard L. Weinstein and Terry L. Clower, "The Economic and Fiscal Impacts of Devon Energy in Denton, Tarrant, and Wise Counties," (Denton: University of North Texas, 2004), 3–7.
5. Weinstein and Clower, 3–7.
6. Weinstein and Clower, 3–7.
7. Bernard L. Weinstein and Terry L. Clower, "The Economic and Fiscal Impacts of Devon Energy Corporation in the Barnett Shale of North Texas: An Update," (Denton: University of North Texas, 2006), 5, 12–13. Of course, rising population and personal income growth was not solely the result of Devon's operations. The nine counties the authors considered were the original three—Denton, Tarrant, and Wise—and Johnson, Parker, Palo Pinto, Erath, Hood, and Hill Counties.
8. David Watts, interview by Diana Davids Hinton, July 20, 2009, Midland, Texas; Jerry Daniel Reed, "An Oily Future: Thousands Hired Over Past Year in Oil and Gas Industry," *Abilene Reporter-News*, April 10, 2005; Alex Branch, "Gas Boom's Dirty Work Provides Life of Peril and Profits," *Midland Reporter-Telegram*, May 28, 2006.
9. Vicki Vaughan, "Rig Companies Cashing In on Exploration Boom,"

Midland Reporter-Telegram, September 17, 2006; Branch, "Gas Boom's Dirty Work;" Terry Mallozzi, "Industry Making Strides in Addressing Challenge of Hiring, Retaining Quality Workers," *Fort Worth Basin Oil and Gas Magazine*, August 2008, https://web.archive.org/web/20110519083630/http://fwbog.com/index.php?page=article&article=27.

10. Galen Scott, "Natural Gas Worth More Than Face Value," *Weatherford Democrat*, September 25, 2006; Watts, interview; Holli Estridge, "Labor Shortage Slams Gas Play," *Dallas Business Journal*, August 3, 2007. It is difficult to determine the extent to which the need for workers benefited women and minorities, in part because industry organizations like the Independent Petroleum Association of America (IPAA) and the Texas Alliance of Energy Producers do not seem to have collected such information, in part because the boom fell between census years; John-Laurent Tronche, "Barnett Bounty Slow to Reach Minorities, Women," *Fort Worth Business Press*, June 23, 2008. It is likely that white males far outnumbered female and minority workers in drilling and service jobs. However, it is likely that women and minorities did benefit substantially from the myriad of opportunities in other sectors growing form the boom—retailing, construction, real estate, and financial services, for example.
11. Pamela Perceval, "After 30 Years in the Energy Industry, Mike Richey is Enjoying Going from 'Bust' to 'Boom,'" *Fort Worth Basin Oil & Gas Magazine*, September 2008, https://web.archive.org/web/20110519083424/http://fwbog.com/index.php?page=article&article=38; Peggy Heinkel-Wolfe, "Energy Boom Spurs New NCTC Program," *Denton Record-Chronicle*, December 7, 2007; Ken Morgan, interview by Diana Davids Hinton, July 30, 2009, Fort Worth, Texas; John-Laurent Tronche, "Energy Academy, Industry Bridging Workforce Gap," *Fort Worth Business Press*, May 24, 2010; Laurie Fox, "Barnett Shale Gas Field Paying Off," *Dallas Morning News*, May 17, 2007; Robert Francis, "Fort Worth Barnett Shale Job Fair a Big Success," *Fort Worth Business Press*, December 10, 2007; Estridge, "Labor Shortage Slams Gas Play."
12. Dan Piller, "Gas Field Is Hot, But Rigs Are Sparse," *Fort Worth Star-Telegram*, September 12, 2005; Elizabeth Souder, "They Know the Drill," *Dallas Morning News*, January 25, 2006; Clayton J. Hein, "Trickle-down Effect Success in Barnett Shale Helping Local Economy," *Wichita Falls Times Record News*, April 16, 2006; Elizabeth Souder, "Barnett Shale Producer Cuts Back Operations," *Denton Record-Chronicle*, October 25, 2006.

13. For a very helpful introduction to this subject with specific reference to the Barnett play, see J. Zack Burt, "Playing the 'Wild Card' in the High-Stakes Game of Urban Drilling: Unconscionability in the Early Barnett Shale Gas Leases," *Texas Wesleyan Law Review*, Fall 2008, 1–30.
14. Searcy, interview; Lowe, interview; Brogdon, interview.
15. Searcy, interview; Brogdon, interview.
16. Searcy, interview; Brogdon, interview; Editorial, "Drilling Disquiet," *Fort Worth Star-Telegram*, April 15, 2002, http://nl.newsbank.com/nl-search/we/Archives; Araceli Arreola, "Haslet Adds Pipeline Inspector," *Fort Worth Star-Telegram*, June 15, 2002; Tanya Eiserer, Jay Parsons, and Tiara Ellis, "Deadly Well Blast Upsets Neighbors," *Dallas Morning News*, April 22, 2006.
17. Kevin Krause, "N. Texas Sweet Spot: Thousands Cashing in on Demand for Natural Gas," *Dallas Morning News*, August 17, 2003.
18. Clifford Krauss, "Fort Worth Copes with Barnett Shale-Fueled Drilling Boom," *Midland Reporter-Telegram*, November 5, 2006.
19. Eric Aasen, "Gas Wells to be Drilled in Irving," *Dallas Morning News*, June 30, 2006; Betty Dillard, "City Parks Benefit from Barnett Shale Windfall," *Fort Worth Business Press*, January 22, 2007.
20. "The Trinity Trail: A Park, or not a Park?," *Fort Worth Star-Telegram*, September 28, 2007; Bud Kennedy, "Here's the Clear-Cut Truth: We're Better Off with Gas Well," *Fort Worth Star-Telegram*, September 9, 2007; Mike Lee, "City Seeks Dismissal of Trinity Trail Drilling Case," *Fort Worth Star-Telegram*, November 27, 2007; Mike Lee, "Judge Throws Out Suit to Stop Gas Drilling Near Trinity Trails," *Fort Worth Star-Telegram*, March 14, 2008. The site is not visible from directly across the river.
21. "Texas Industry Tempers Local Rules," *American Oil & Gas Reporter* (February 2002): 36.
22. Brogdon, interview.
23. Mike Moncrief, interview by Diana Davids Hinton, February 28, 2013, Fort Worth, Texas.
24. Peggy Williams, "The Barnett Shale," *Oil and Gas Investor*, March 2002, 41.
25. Don and Debora Young, interview by Diana Davids Hinton, July 28, 2010, Fort Worth, Texas.
26. Young interview.
27. Young interview.
28. Don Young, "Lipstick on a Poison Pig," *Fort Worth Weekly*, April 26, 2006; Robert Francis, "Drilling Task Force Reports to City, Suggest Ordinance,"

Fort Worth Business Press, April 10, 2006; Aleshia Howe, "Council Approves 600–foot Gap Between Gas Wells, Neighborhoods," *Fort Worth Business Press*, June 19, 2006; Jenny Eure, "Gas Wells Fuel Real Estate Debate," *Fort Worth Business Press*, October 9, 2006. Denton County established an Oil and Gas Task Force in 2004.

29. City of Fort Worth, Ordinance Number 16986–06–2006: "An Ordinance Amending the Code of Ordinances of the City of Fort Worth, By Amending Article 11 of Chapter 15, 'Gas,' entitled, 'Gas Drilling and Production,' Regulating the Drilling and Production of Gas Wells Within the City to Provide Revised Regulations Regarding Distance, Noise And Technical Provisions; Providing That This Ordinance Shall Be Cumulative Of All Ordinances; Providing A Savings Clause; Providing A Severability Clause; Providing a Penalty Clause, Providing For Publication; And Naming An Effective Date."
30. Matt Smith, "Business Receives 2–Year Tax Reduction," *Cleburne Times-Review*, January 14, 2010; Rob Fraser, "Energy Driving Cleburne Growth," *Cleburne Times-Review*, June 6, 2006; Betty Dillard, "Oil Roads Lead to Cleburne," *Fort Worth Business Press*, November 19, 2007; David Watts, interview; Jerry Cash, interview by Diana Davids Hinton, July 29, 2010, Cleburne, Texas.
31. Fraser, "Energy Driving Cleburne Growth;" Watts, interview; Dillard, "Oil Roads Lead to Cleburne."
32. The Perryman Group, *Long-Term Effect of the Barnett Shale* (np: The Perryman Group, 2007), 133; City of Cleburne, "Johnson County Unemployment Rate Comparison 2008 thru 2010," *Monthly Economic Development Trends Report*, in City of Cleburne, July 16, 2010, 4; Cash, interview; Watts, interview; David H. Arrington, interview by Diana Davids Hinton, October 16, 2009, Midland, Texas.
33. Arrington, interview; Cash, interview; Perryman Group, *Long Term Effect*, 113. The Perryman Group calculated that in 2007 royalty and lease payments amounted to over $189 million in Johnson County.
34. City of Cleburne, "State of the City Report" (2010), 15; Matt Smith, "Former Sports Complex Called Unsafe," *Cleburne Times-Review*, March 2, 2007; "Drilling Activity Affecting HOT," *Cleburne Times-Review*, December 20, 2009; Matt Smith, "Local Stimulus Package Announced at Workshop," *Cleburne Times-Review*, October 18, 2009; Michael O'Connor, "Foundation Provides Solid Base for City Business," *Cleburne Times-Review*, November 11, 2008.

35. Roger Harmon, interview by Diana Davids Hinton, July 29, 2010, Cleburne, Texas; Misty Shultz, "Energy Company Gives $106,000," *Cleburne Times-Review*, August 28, 2007; Lela Jobe, "Companies Contribute $827,000 for Roads," *Cleburne Times-Review*, December 30, 2007; Misty Shultz, "General Fund $1.5 Million Higher Than Fiscal 2005," *Cleburne Times-Review*, June 20, 2007. A later observer of the Eagle Ford Shale boom in rural South Texas estimated that it took 1200 loaded semi-trailers to put in a well, 350 trucks to maintain it for a year, and 1000 trucks to re-frack it; John Mangalonzo, "Big Country's Cline Shale Will Bring Dramatic Changes, Summit Told," *Abilene Reporter-News*, February 8, 2013. It is worth noting that when Johnson County gas well drilling fell off, so did company contributions.
36. Watts, interview.
37. *Cleburne Times-Review*, "Letter to the Editor," December 16, 2007.

CHAPTER 5

1. Robert Francis, "Barnett Shale Helping Sell Million-Dollar Homes in Area," *Fort Worth Business Press*, January 25, 2008.
2. The Perryman Group, "An Enduring Resource: A Perspective on the Past, Present and Future Contribution of the Barnett Shale to the Economy of Fort Worth and the Surrounding Area," (Waco: The Perryman Group, 2009), 73–74; Betty Dillard, "New Projects Will Change Face of Fort Worth in 2008," *Fort Worth Business Press*, December 31, 2007; Ed Ireland, "Barnett Shale Spurs New Jobs and Business Growth," *Fort Worth Business Press*, February 11, 2008.
3. David Wethe, "Chesapeake Buys Sundance Lease," *Fort Worth Star-Telegram*, May 3, 2008; Robert Francis, "Barnett Shale Casts Shadow over City," *Fort Worth Business Press*, January 1, 2007; Betty Dillard, "New Projects will Change the Face of Fort Worth in 2008," *Fort Worth Business Press*, December 31, 2007; Robert Francis, "Barnett Shale Helping Sell Million-Dollar Homes in Area," *Fort Worth Business Press*, January 28, 2008.
4. John Burnett, "Urban Gas Drilling Causes Backlash in Boomtown," *NPR*, August 21, 2008; Peter Gorman, "Blue-Collar Paradise," *Fort Worth Weekly*, March 12, 2008.
5. *Fort Worth Star-Telegram*, September 28, 2007.

6. Ed Ireland, interview by Diana Davids Hinton, July 31, 2009, Fort Worth, Texas; Galen Scott, "Barnett Shale Leaders Launch Web Site," *Weatherford Democrat*, November 6, 2007; Mike Moncrief, interview.
7. *Fort Worth Star-Telegram*, March 23, 2008.
8. Ben Casselman, "Chesapeake Energy to Deliver 'News,' Too," *Wall Street Journal*, July 22, 2008; John-Laurent Tronche, "Chesapeake Hires North Texas Newsman to Provide In-Depth Shale Education," *Fort Worth Business Press*, July 10, 2008; Mitchell Schnurman, "Working Overtime on Hearts and Minds," *Fort Worth Star-Telegram*, June 22, 2008.
9. Reese Gordon, "Date to Start Gas Drilling Up in Air," The *[Texas Christian University] Daily Skiff*, January 24, 2008; David Wethe, "Chesapeake Buys Sundance Lease," *Fort Worth Star-Telegram*, May 3, 2008; John-Laurent Tronche, "TCU Alumnus Key to Drilling Plan," *Fort Worth Business Press*, April 13, 2009; Nowell Donovan, interview by Diana Davids Hinton, June 23, 2014, by telephone.
10. Robert Francis, "Chesapeake Lands D/FW Drilling Contract," *Fort Worth Business Press*, August 7, 2006; Robert Francis, "In Chesapeake Deal, D/FW Airport Finds 'Sweet' Gas," *Fort Worth Business Press*, October 16, 2006.
11. L. Decker Dawson, C. Ray Tobias, and Steve C. Jumper, interview by Diana Davids Hinton, July 6, 2009, Midland, Texas.
12. Robert Francis, "Chesapeake Makes Hard Landing at D/FW Airport," *Fort Worth Business Press*, May 28, 2007; Robert Francis, "Chesapeake Production Takes Off at D/FW Airport," *Fort Worth Business Press*, November 5, 2007; John Armistead, "Chesapeake at D/FW Airport Initial Results Positive for $186 Million Gamble," *Fort Worth Business Press*, March 3, 2008.
13. Rick Holden, interview by Diana Davids Hinton, March 1, 2013, Cleburne, Texas.
14. John-Laurent Tronche, "Cities Spend Natural Gas Cash on One-Time Expenditures," *Fort Worth Business Press*, July 13, 2009; "Mineral Rights Manna," *Fort Worth Weekly*, March 14, 2007; Mike Lee, "Plans for Natural Gas Windfall Finalized," *Fort Worth Star-Telegram*, January 9, 2008; Mike Moncrief, interview.
15. "Mineral-Rights Manna," *Fort Worth Weekly*, March 14, 2007.
16. Aleshia Howe, "Barnett Shale Now a Factor in Real Estate Transactions," *Fort Worth Business Press*, May 12, 2008. Since bonuses were figured by acre, some canny lot owners insisted that their acreage be figured to include area extending to the middle of the street bordering their lot; "N.

Benbrook Neighborhoods Get $15,000 Deal," *Fort Worth Star-Telegram*, January 22, 2008.

17. John-Laurent Tronche, "Landmen, County Disagree on How to Solve Courthouse Congestion," *Fort Worth Business Press*, May 12, 2008.
18. "East Fort Worth Group Swings for the Fences: $30,000 and 30%," *Fort Worth Star-Telegram*, January 17, 2008; "360 Northwest Coalition Inks $23,500 Deal with Chesapeake," *Fort Worth Star-Telegram*, July 2008.
19. "An Early Look at the Mansfield South Deal," *Fort Worth Star-Telegram*, April 17, 2008; "Overton, Southwest Fort Worth Alliance Groups Endorse Leases," *Fort Worth Star-Telegram*, September 8, 2008; "Mont Del Lease: Near-Record Bonus, HOA Cash, and a Bench," *Fort Worth Star-Telegram*, June 26, 2008; Mike Moncrief, interview.
20. "Arlington's Woodland West: Some Happy, Some Not," *Fort Worth Star-Telegram*, January 9, 2008.
21. "Class-Action Status Sought for Glencrest Lawsuit," *Fort Worth Star-Telegram*, November 1, 2007; "Briscoe's Glencrest Resources Paying Bonuses Now," *Fort Worth Star-Telegram*, May 27, 2008; "Dallas Law Firms Invite Reports of Misrepresentation," *Fort Worth Star-Telegram*, March 27, 2008.
22. "More Southeast Arlington Leasing," *Fort Worth Star-Telegram*, January 24, 2008; Mike Lee, "Neighborhood Group, Driller Spar over Site," *Fort Worth Star-Telegram*, October 13, 2007; "Brentwood/Oak Hills Relieved over Drilling Site," *Fort Worth Star-Telegram*, February 11, 2008. Paloma sold its leases to Chesapeake, which decided it would not use the controversial drilling site.
23. David Wethe, "New Ordinance has Chesapeake Looking Elsewhere," *Fort Worth Star-Telegram*, May 31, 2008; "Chesapeake Takes a Breather in Southlake," *Fort Worth Star-Telegram*, July 29, 2008.
24. Sam Fletcher, "Energy Prices Continue upward Climb," *Oil & Gas Journal*, May 7, 2008, http://www.ogj.com/articles/2008/05/market-watch-energy-prices-continue-upward-climb.html; Sam Fletcher, "Crude Price Makes Record Jump Past $138/bbl," *Oil & Gas Journal*, June 9, 2008, http://www.ogj.com/articles/2008/06/market-watch-crude-price-makes-record-jump-past-138–bbl.html; Sam Fletcher, "Energy Prices Register New Intraday Highs," *Oil & Gas Journal*, July 14, 2008, http://www.ogj.com/articles/2008/07/market-watch-energy-prices-register-new-intraday-highs.html; Nick Snow, "Cause of 2008's Crude Oil Price Surge Remains Elusive, CFTC Says," *Oil & Gas Journal*, April 9, 2009,

http://www.ogj.com/articles/2009/04/cause-of-2008s-crude-oil-price-surge-remains-elusive-cftc-says.html; Mitchell Schnurman, "The Bubble Bursts, and the Big Bonuses Vanish," *Fort Worth Star-Telegram*, October 26, 2008.

25. John-Laurent Tronche, "Barnett Shale Operators Stung in Financial Hornets Nest," *Fort Worth Business Press*, October 6, 2008.
26. Jack Z. Smith, "Low Natural Gas Prices Hurting Many North Texans," *Fort Worth Star-Telegram*, August 23, 2009; Candace Carlisle, "Lower Gas Prices Hurt Area Economy," *Denton Record-Chronicle*, September 8, 2009; John-Laurent Tronche, "Quicksilver Cuts Budget, Drops Barnett Shale Rigs," *Fort Worth Business Press*, October 8, 2008; "Devon Chief Says Drilling Slowdown Will Continue," *Fort Worth Star-Telegram*, June 3, 2009, http://nl.newsbank.com/nl-search/we/Archives; Makaila Adams, "Chesapeake Shares Rise on Take Over Talk," *Oil and Gas Financial Journal*, November 24, 2008, http://www.ogfj.com/articles/2008/11/chesapeake-shares-rise-on-takeover-talk.html.
27. Mitchell Schnurman, "The Bubble Bursts, and the Big Bonuses Vanish," *Fort Worth Star-Telegram*, October 26, 2008; Elizabeth Souder and Maurice Richter, "As the Barnett Shale's Production Peaks, Economic Factors Weigh Heavily on its Future," *Dallas Morning News*, October 26, 2008.
28. Souder and Richter, "Barnett Shale's Production Peaks"; "Colleyville Residents Fault Council for Approving Gas Deal," *Fort Worth Star-Telegram*, October 22, 2008; John-Laurent Tronche, "Signing Bonuses Disappear as Operators Adjust to Tough Times," *Fort Worth Business Press*, November 3, 2008.
29. John-Laurent Tronche, "Landmen Rush Fades as Leasing Activity Slows," *Fort Worth Business Press*, November 17, 2008; John-Laurent Tronche, "Pipeline Infrastructure Necessitates Work as Layoffs Continue," *Fort Worth Business Press*, April 27, 2009; Robert Francis, "Barnett Shale Rig Count Drops More than 50 Percent," *Fort Worth Business Press*, March 3, 2009; Peggy Heinkel-Wolfe, "Barnett Shale-Related Jobs Decline," *Denton Record-Chronicle*, March 12, 2009; John-Laurent Tronche, "Beyond the Barnett: Energy Industry Eyeing Other Shales," *Fort Worth Business Press*, April 13, 2009; Aleshia Howe, "Commercial Foreclosures Increase in Dallas—Fort Worth," *Fort Worth Business Press*, December 15, 2008; Aleshia Howe, "Reality in Forecast for Tarrant Commercial Real Estate," *Fort Worth Business Press*, January 26, 2009.

30. Aman Batheja, “Gas Funds May Help Ease Area Shortfalls,” *Fort Worth Star-Telegram*, August 29, 2010; Holden, interview.
31. Batheja, “Gas Funds May Help;” Leslie Wimmer, “Fort Worth Budget Gap Grows,” *Fort Worth Business Press*, May 31, 2010.
32. “Denbury to Sell Barnett Shale Assets,” *Dallas Business Journal*, May 14, 2009; John-Laurent Tronche, “Beyond the Barnett: Energy Industry Eyeing Other Shales,” *Fort Worth Business Press*, April 13, 2009; Elizabeth Souder, “Quicksilver Resources Sells Some Barnett Shale Leases for $280 M,” *Dallas Morning News*, May 18, 2009; Bob Cox, “XTO Shareholders Voting on Sale to Exxon,” *Fort Worth Star-Telegram*, June 25, 2010.

CHAPTER 6

1. Gene Theodori, “Public Opinion Research on Urban Gas Drillers,” Center for Rural Studies: Research & Outreach, Sam Houston State University, 2009; Dan Williams, interview by Diana Davids Hinton, July 30, 2010, Fort Worth, TX; Nowell Donovan, interview by Diana Davids Hinton, June 23, 2014, telephone.
2. Mike Lee, “City Residents Voice Drilling Concerns,” *Fort Worth Star-Telegram*, September 29, 2007; Mike Lee, “Neighborhood Group, Driller Spar Over Site,” *Fort Worth Star-Telegram*, October 13, 2007; Bud Kennedy, “If You Don’t Follow Rules, Your Sign Becomes the City’s,” *Fort Worth Star-Telegram*, September 12, 2007.
3. Peter Gorman, “Hot Water: Eastside Activists Join a Growing Line of Folks Worried about Gas-Drilling Waste Products,” *Fort Worth Weekly*, November 28, 2007.
4. Bob Ray Sanders, “That Upset Feeling—Was It Something We Swallowed? Barnett Shale,” *Fort Worth Star-Telegram*, December 9, 2007.
5. Peter Gorman, “Perilous Profits: Tempers and Dangers Are Ramping Up in the Gas Field We Call Home,” *Fort Worth Weekly*, March 28, 2007; Peter Gorman, “Digging Deeper: Urban-Gas-Well Activists Face an Uphill Battle Now that Real Money’s on the Table,” *Fort Worth Weekly*, November 14, 2007; Judy Alter, interview by Diana Davids Hinton, July 27, 2009, Fort Worth, TX.
6. Lee, “City Residents Voice Drilling Concerns”; Mike Lee, “Mayor Says Oil

and Gas Income is Not a Conflict," *Fort Worth Star-Telegram*, September 29, 2007; Mike Moncrief, interview.

7. Will Brackett, "New Task Force Set to Study Barnett Shale Gas Well Ordinance," *Fort Worth Business Press*, February 15, 2008; Jeff Prince, "Just Say Whoa," *Fort Worth Weekly*, July 30, 2008.
8. Prince, "Just Say Whoa"; John-Laurent Tronche, "Gas Drilling Moratorium Rally Calls for Halt to Natural Gas Drilling," *Fort Worth Business Press*, August 14, 2008.
9. Mike Lee, "On Carter Avenue, Residents are Learning What Happens When a Gas Company Wants to Build a Pipeline," *Fort Worth Star-Telegram*, June 6, 2008; Jeff Prince, "He Won't Pipe Down," *Fort Worth Weekly*, January 13, 2010. On the issue of eminent domain in Texas law, see Michael B. Lee, "Pipelines and Eminent Domain in Texas," a paper presented to the Seventh Annual Eminent Domain Super Conference: Current Texas Law and Policy, Eminent Domain Institute, Austin, Texas, February 2008.
10. Mike Lee, "Update on Carter Avenue," *Fort Worth Star-Telegram*, July 29, 2008; "Gas Company Tries to Condemn Land for Pipeline," *North Texas Daily*, August 2, 2008; Mike Lee, "Carter Avenue, Update II," *Fort Worth Star-Telegram*, July 31, 2008; Mike Lee, "Carter Avenue Update," *Fort Worth Star-Telegram*, August 21, 2008; Prince, "He Won't Pipe Down."
11. Prince, "He Won't Pipe Down"; Mike Lee, "State Highway Department Says Carter Avenue Pipeline Might Run Along I-30," *Fort Worth Star-Telegram*, November 24, 2009; Mike Lee, "Pipeline Bill Would Allow Pipelines in Freeway Right of Way," *Fort Worth Star-Telegram*, February 6, 2009; Mike Lee, "Carter Avenue Landowner Loses Round in Court over Gas Pipeline," *Fort Worth Star-Telegram*, March 5, 2010; Mike Lee, "Usable Route for Gas Pipeline Found Along Interstate I-30," *Fort Worth Star-Telegram*, March 12, 2010; Peter Gorman, "Shift in the Shale," *Fort Worth Weekly*, December 16, 2009.
12. Mike Lee, "Gas Drilling Committee Gets Down to Nitty-Gritty," *Fort Worth Star-Telegram*, April 15, 2008; Peter Gorman, "A New Drill," *Fort Worth Weekly*, December 17, 2008.
13. Lee, "Gas Drilling Committee Gets Down to Nitty-Gritty"; Gorman, "A New Drill."
14. Lee, "Gas Drilling Committee Gets Down to Nitty-Gritty"; Gorman, "A New Drill."; City of Fort Worth, Ordinance Number 18399–12–2008: "An

Ordinance Amending the Code of Ordinances of the City of Fort Worth by Amending Article II of Chapter 15, 'Gas,' Entitled, 'Gas Drilling and Production,' Regulating the Drilling and Production of Gas Wells Within the City to Provide Revised Regulations Regarding Distance, Noise, Gas Pipelines, and Technical Provisions; Providing that this Ordinance Shall Be Cumulative of All Ordinances; Providing a Savings Clause; Providing a Severability Clause; Providing a Penalty Clause; Providing for Publication; and Naming an Effective Date"; John-Laurent Tronche, "More Natural Gas Drilling Ordinance Changes to Come," *Fort Worth Business Press*, December 15, 2008.

15. John-Laurent Tronche, "Bedford Gas Rules Draw Much Emotion, Little Action," *Fort Worth Business Press*, February 25, 2008; Pablo Lastra, "Canary in the Coal Mine Gas Field," *Fort Worth Weekly*, August 27, 2008; Mike Lee, "Tarrant Urban Drilling: 'The Early Stuff's Done,'" *Fort Worth Star-Telegram*, June 27, 2010; David Wethe, "New Ordinance Has Chesapeake Looking Elsewhere," *Fort Worth Star-Telegram*, May 31, 2008; Jim Fuquay, "Chesapeake Takes a Breather in Southlake," *Fort Worth Star-Telegram*, July 29, 2008.
16. Al Armendariz, "Emissions from Natural Gas Production in the Barnett Shale Area and Opportunities for Cost-Effective Improvements," prepared for Ramon Alvarez, Environmental Defense Fund, January 26, 2009, http://www.edf.org/sites/default/files/9235_Barnett_Shale_Report.pdf; Mike Lee, "Conflict is Nothing New for EPA Official," *Fort Worth Star-Telegram*, February 21, 2010.
17. Galen Scott, "Regional Focus on Air Quality," *Weatherford Democrat*, June 22, 2007; Eric Griffey, "The Opacity of Hope," *Fort Worth Weekly*, December 16, 2009; Pamela Percival, "Infrared Cameras Used to Locate 'Fugitive' Vent Gas," *Fort Worth Basin Oil & Gas Magazine*, February 2009.
18. Chris Vaughn, "Air Quality Tests Raise Questions about Natural Gas Wells in the Barnett Shale," *Fort Worth Star-Telegram*, October 3, 2009; Wolf Eagle Environmental Engineers and Consultants, "Deborah Rogers/HDLA, LLC (d.b.a. Deborah's Farmstead) Ambient Air Monitoring Analysis Project: Final Report," May 25, 2009; Sarah Perry, "Regulation Stymied by Limits, Differences," *Denton Record-Chronicle*, March 30, 2011.
19. Perry, "Regulation Stymied"; Vaughn, "Air Quality Tests"; Peter Gorman, "The Barnett Keeps Bubbling," *Fort Worth Weekly*, August 18, 2010.
20. Industrial Hygiene and Safety Technology, Inc., "Review of Ambient Air

Monitoring Project and Related Communications Concerning Gas Well Emissions Provided for the City of Fort Worth Environmental Management Department," August 24, 2009, 3–4, 6–8.

21. Ed Ireland, "News from the Barnett Shale Energy Education Council," *Fort Worth Basin Oil & Gas Magazine*, December 2009; Ed Ireland, "Air Pollution Issues in the Barnett Shale," Barnett Shale Energy Education Council, December 3, 2009, http://www.bseec.org/content/benzene-issues-barnett-shale.
22. Peter Gorman, "Sacrificed to the Shale," *Fort Worth Weekly*, October 14, 2009. The satellite company wanted the town's name to be spelled with capital letters: DISH. I have chosen the more conventional format for a name.
23. Peggy Heinkel-Wolfe, "Energy Officials Dish Out Answers for Area Residents," *Denton Record-Chronicle*, November 16, 2006.
24. Gorman, "Sacrificed to the Shale."
25. Gorman, "Sacrificed to the Shale."
26. Wolf Eagle Environmental, "Town of DISH, Texas Ambient Air Monitoring Analysis Final Report," September 15, 2009, 2. The company had dropped "Engineers and Consultants," perhaps because its critics pointed out none of its members were engineers.
27. Wolf Eagle Environmental, 3–4, 6–9.
28. Wolf Eagle Environmental, 9.
29. Wilma Subra, "Evaluation of Town of DISH, Texas Ambient Air Monitoring Analysis," October 6, 2009, https://www.earthworksaction.org/library/detail/evaluation_of_town_of_DISH_texas_ambient_air_monitoring_analysis#.WhtCnLT828V.
30. Wilma Subra, "Health Survey Results of Current and Former DISH/Clark, Texas Residents," December 17, 2009, 3, https://www.earthworksaction.org/library/detail/health_survey_results_of_current_and_former_dish_clark_texas_residents/#.WhtC_bT828U.
31. Subra, 3, 11–12.
32. Subra, 7, 12–13.
33. Shannon Ethridge, "Health Effects Review of Ambient Air Monitoring Data Collected by Wolf Eagle Environmental Engineers and Consultants for DISH, TX.," interoffice memorandum, October 27, 2009, *Summary of TCEQ Actions Relating to the Barnett Shale, Attachment 8*, November 13, 2009, http://www.nctcog.org/trans/air/TCEQHealthEffectsEvaluationforDISHTX.pdf;

Texas Commission on Environmental Quality, "Responses to Representative Lon Burnam's Oil and Gas Questions," October 1, 2009, *Summary of TCEQ Actions Relating to the Barnett Shale, Attachment* 9, November 13, 2009, https://www.oilandgaslawyerblog.com/files/2015/02/TCEQ-Summary-with-Burnam-QA1.pdf; For a trenchant and highly readable critique of both Wolf Eagle and Subra's work, see Jeff R. Bowman's "Air Quality in the Barnett Shale—A Thirty-one Part Scientific Study," a series of postings from December 1, 2010 to March 26, 2011, http://wackyworldofwaste.blogspot.com/2010/12/, http://wackyworldofwaste.blogspot.com/2011/01/, http://wackyworldofwaste.blogspot.com/2011/02/, http://wackyworldofwaste.blogspot.com/2011/03/

34. Peggy Heinkel-Wolfe, "High Emissions Levels Recorded Outside DISH TCEQ Still Analyzing Data on Toxic Compounds Found During Site Visits," *Denton Record-Chronicle*, November 24, 2009; Peggy Heinkel-Wolfe, "Dish Air Study Detects Pollution," *Denton Record-Chronicle*, October 11, 2009.
35. Gorman, "Sacrificed to the Shale"; Mark Bauer, "Gas Wells Emissions Could Cause Health Concerns, Study Says," *The Shorthorn*, December 1, 2009, http://www.theshorthorn.com/news/gas-wells-emissions-could-cause-health-concerns-study-says/article_7bf864cc-b46c-5375-b1c6-33866fc14da0.html.
36. Jeff Prince, "The Quality of the Test," *Fort Worth Weekly*, January 27, 2010, https://www.fwweekly.com/2010/01/27/the-quality-of-the-test/; Gorman, "Sacrificed to the Shale."
37. Ireland, "News from the Barnett Shale Energy Education Council"; Jeff Mosier, "Air Tests at Natural Gas Drilling Sites Fuel Concerns in Barnett Shale," *Dallas Morning News*, December 4, 2009, https://www.dallasnews.com/news/texas/2009/12/04/20091203–Air-tests-at-natural-gas-drilling-3764; Jeff Mosier, "Fort Worth Council Members Ask State to Test City Air for Natural Gas Emissions," *Dallas Morning News*, December 9, 2009; John-Laurent Tronche, "Area Debates Natural Gas Emissions," *Fort Worth Business Press*, December 14, 2009.
38. Mike Lee and Aman Batheja, "Audit: Agency Gave Inaccurate Air Pollution Test Results to Fort Worth," *Fort Worth Star-Telegram*, May 27, 2010; Mike Lee, "Tests Find High Benzene Levels at Two More Sites," *Fort Worth Star-Telegram*, June 2010.
39. Mike Lee and Jeff Claassen, "Audit Takes Commission to Task," *Fort Worth*

Star-Telegram, October 1, 2007; "Texas Railroad Commission to Open Fort Worth Office," *Fort Worth Basin Oil & Gas Magazine*, October 2009, https://web.archive.org/web/20110519083411/http://fwbog.com/index.php?page=article&article=37; Asher Price, "Environmental, Oil Agencies Told to Be Tougher on Violators," *Austin American-Statesman*, November 18, 2010.

40. William R. Childs, *The Texas Railroad Commission: Understanding Regulation in America to the Mid-twentieth Century* (College Station: Texas A&M University Press, 2005), 217–24.
41. Jack Z. Smith, "EPA to Seek Public Comment on Hydraulic Fracturing in Fort Worth," *Fort Worth Star-Telegram*, June 29, 2010; Griffey, "The Opacity of Hope"; Mike Lee, "Conflict in Nothing New for EPA Official," *Fort Worth Star-Telegram*, February 21, 2010

CHAPTER 7

1. Elizabeth Campbell and Bill Teeter, "Drought, Population Growth Contribute to Lower Water in Regional Aquifer," *Fort Worth Star-Telegram*, August 26, 2006; R. W. Harden & Associates, Inc., "Northern Trinity/Woodbine Groundwater Availability Model Assessment of Groundwater Use in the Northern Trinity Aquifer Due To Urban Growth and Barnett Shale Development," (Austin: Texas Water Development Board, 2007), 2, https://www.twdb.texas.gov/groundwater/models/gam/trnt_n/TRNT_N_Barnett_Shale_Report.pdf; Robert Francis, "Water Pressures Drillers to Form Task Force," *Fort Worth Business Press*, September 18, 2006.
2. Peter Gorman, "Water Foul," *Fort Worth Weekly*, April 30, 2008; Elizabeth Campbell and Aman Batheja, "Denton, Johnson County Residents Blame Drilling Process for Fouled Well Water," *Fort Worth Star-Telegram*, July 1, 2010, http://www.star-telegram.com/living/family/moms/article3825756.html; Jack Z. Smith, "The Barnett Shale Search for Facts on Fracking," *Fort Worth Star-Telegram*, September 4, 2010.
3. Jeff Prince, "We Are Doing It," *Fort Worth Weekly*, March 25, 2009; Sharon Wilson, "Oil and Gas Industry Has Duty to Minimize Air Pollution," *Wise County Messenger*, December 18, 2008; "Bluedaze: Drilling Reform for Texas," December 15, 2007, http://www.texassharon.com/2007/12/15/how-i-became-a-"far-left-radical-with-a-socialist-agenda"-etc/; John-Laurent

Tronche, "New Watchdog Group Aims to Rein In Natural Gas Industry," *Fort Worth Business Press*, March 1, 2010.

4. John-Laurent Tronche, "U. S. Representatives Unveil FRAC Act to Close Halliburton Loophole," *Fort Worth Business Press*, June 9, 2009.
5. John-Laurent Tronche, "House of Reps. Opens Investigation of Hydraulic Fracturing," *Fort Worth Business Press*, February 15, 2010; John-Laurent Tronche, "EPA Initiates Hydraulic Fracturing Study," *Fort Worth Business Press*, March 22, 2010.
6. Blaine D. Edwards, E. James Shepherd, and Nick Deutsch, "Hydraulic Fracturing: Protecting Against Legal and Regulatory Risk," *Oil & Gas Journal*, August 1, 2011, 22–24, 26–28, http://www.ogj.com/articles/print/volume-109/issue-31/general-interest/hydraulic-fracturing-protecting-against.html.
7. Mike Lee, "EPA Still Working on Barnett Shale Air Pollution Problem, Agency Says," *Fort Worth Star-Telegram*, May 11, 2010, http://www.star-telegram.com/living/family/moms/article3825481.html.
8. Jack Z. Smith, "Fort Worth Meeting on Gas Drilling Process Draws Heated Response," *Fort Worth Star-Telegram*, July 8, 2010.
9. Aman Batheja, "A Controversial Look at Drilling," *Fort Worth Star-Telegram*, June 21, 2010.
10. Mike Hale, "The Costs of Natural Gas, Including Flaming Water," *New York Times*, June 20, 2010, http://www.nytimes.com/2010/06/21/arts/television/21gasland.html; Errol Lewis, "Natural Gas, Unnatural Risk," *New York Daily News*, July 25, 2010; Susan Breslow, "'Gasland' Movie Review—and Call to Action," *New York Examiner*, June 18, 2010.
11. Railroad Commission of Texas, Office of the General Counsel, "Oil and Gas Docket No. 7B-0268629: Commission Called Hearing to Consider Whether Operation of the Range Production Company Butler Unit Well No. 1H (RRC ID 253732) and Teal Unit Well No. 1H (RRC ID 253729) in the Newark, East (Barnett Shale) Field, Hood County, Texas are Causing or Contributing to Contamination of Certain Domestic Water Wells in Parker County, Texas," January 19–20, 2011, 2–3, http://www.rrc.texas.gov/media/10502/7b-68629rangepfd-03–11–11–commcalledepa.pdf.
12. Shannon Brushe, "EPA's 'Yee Haw' Moment in Texas," *Energy In Depth*, February 14, 2011, http://energyindepth.org/national/epas-yee-haw-moment-in-texas-2/; Jack Z. Smith, "Flower Mound Scientist Often Plays Key Role in Drilling Controversies," *Fort Worth Star-Telegram*, July 9, 2011.

13. Brushe, "EPA's 'Yee Haw' Moment."
14. Bill Hanna and Jack Z. Smith, "EPA Blames Drilling for Fouled Water at 2 Homes," *Fort Worth Star-Telegram*, December 8, 2010.
15. Railroad Commission of Texas, "Oil and Gas Docket No. 7B-0268629", 3, http://www.rrc.state.tx.us/media/10497/7b-68629–commcalled-epa.pdf; Brushe, "EPA's 'Yee Haw' Moment."
16. Brushe, "EPA's 'Yee Haw' Moment."
17. Chris Hawes, "EPA Acts After Water Contaminated by Drilling in Parker County," in "EPA's 'Yee Haw' Moment;" Bill Hanna and Jack Z. Smith, "Driller Denies Blame for Methane in Parker County Homes' Water," *Fort Worth Star-Telegram*," December 7, 2010.
18. "The Texas Railroad Commission Fiddles While Well Water Burns," *Fort Worth Star-Telegram*, December 9, 2010.
19. "Railroad Commission Fiddles;" Hanna and Smith, "EPA Blames Drilling."
20. Gene Powell, "EPA Wrong—Barnett Shale Gas Not in Water Wells," *Powell Barnett Shale Newsletter*, December 13, 2010. Powell consistently supported industry on Barnett Shale issues.
21. Jack Z. Smith, "Ruling on Contaminants Questioned, Water Well Drillers Not So Sure that Methane in Residents' Groundwater is from Barnett Shale Activity," *Fort Worth Star-Telegram*, December 19, 2010.
22. Jack Z. Smith, "EPA, Range in Fight over Contamination Claim," *Fort Worth Star-Telegram*, January 9, 2011.
23. Gene Powell, "No Credible Evidence Supports EPA Order Against Range," *Powell Barnett Shale Newsletter*, February 11, 2011; Jack Z. Smith, "EPA Official is Grilled in Range Water Well Case," *Fort Worth Star-Telegram*, February 11, 2011.
24. Railroad Commission of Texas, Oil and Gas Docket No. 7B-0268629, 8–9, http://www.rrc.state.tx.us/media/10497/7b-68629–commcalled-epa.pdf.
25. Railroad Commission of Texas, 6–7, 10–11.
26. Ramit Plushnick-Masti, "Altering the Flow: EPA Changed Course After Oil Company Protested in Long-Standing Legal Battle over Gas-Tainted Water Supply," *Weatherford Democrat*, January 17, 2013; Randy Lee Loftis, "EPA to Expand Dallas-Fort Worth Smog-Violations Area," *Dallas Morning News*, December 9, 2011, https://www.dallasnews.com/news/news/2011/12/09/epa-to-expand-dallas-fort-worth-smog-violation-area.
27. Jack Z. Smith, "Judge Declines to Dismiss EPA Order Against Range Resources," *Fort Worth Star-Telegram*, June 21, 2011; Lance Winter, "Couple

Seeks $6.5 Million in Damages for Water Well Contamination," *Weatherford Telegram*, June 24, 2011; Jack Z. Smith, "Flower Mound Scientist Often Plays Key Role in Drilling Controversies," *Fort Worth Star-Telegram*, July 9, 2011.

28. Jack Z. Smith, "EPA E-Mail Stirs Call for Agency Documents," *Fort Worth Star-Telegram*, February 22, 2011; Mike Soraghan, "Texas EPA Official's E-Mails Show Federal-State Tension over Sanctions on Natural Gas Drilling," *New York Times*, February 11, 2011, http://www.nytimes.com/gwire/2011/02/11/11greenwire-texas-epa-officials-e-mails-show-federal-state-63373.html?pagewanted=all; Dave Michaels, "Chairman Comes to Defense of Oil and Gas Drillers," *Dallas Morning News*, July 16, 2011.
29. Dave Michaels, "EPA Takes Aim at Gas Well Pollution," *Dallas Morning News*, July 29, 2011; Deborah Solomon and Russell Gold, "EPA Ties Fracking, Pollution," *Wall Street Journal*, December 9, 2011, https://www.wsj.com/articles/SB10001424052970203501304577086472373346232; Tom Fowler, "Fracturing Report Focuses on Wells and Air Emissions," *Midland Reporter-Telegram*, August 14, 2011. As in Parker County, a shallow gas formation at Pavillion could as readily have been the source of water contamination as industry operations.
30. Ramit Plushnick-Masti, "Altering the Flow"; Barry Shlachter, "EPA Drops Action Against Range Resources over Parker County Water Wells," *Fort Worth Star-Telegram*, March 31, 2012, http://www.star-telegram.com/living/family/moms/article3830897.html; Sharon Wilson (TXsharon), "Groups Urge Investigation of EPA Actions in Texas Water Contamination Case," February 11, 2013, http://www.texassharon.com/2013/02/11/groups-urge-investigation-of-epa-actions-in-texas-water-contamination-case/.
31. James M. Inhofe to Lisa Jackson, April 25, 2012, https://www.epw.senate.gov/public/_cache/files/7/d/7d4b80e5–a4f8–42c0–8c62–8cec8fd5e614/01AFD79733D77F24A71FEF9DAFCCB056.regionvilettertojackson.pdf.
32. Inhofe to Lisa Jackson.
33. Nick Snow, "EPA Region 6 Chief Resigns over 2010 'Crucifixion' Remark," *Oil & Gas Journal*, May 1, 2012, http://www.ogj.com/articles/print/vol-110/issue-5a/general-interest/epa-region-6–chief-resigns.html; Steve Everley, "Update XX, EPA Official: 'Crucify' Operators to 'Make Examples' of Them," *Energy In Depth*, February 22, 2013, http://energyindepth.org/national/epa-regional-admin-crucify-operators-to-make-examples/;

John M. Broder, "E.P.A. Official in Texas Quits over 'Crucify' Video," *New York Times*, May 1, 2012, http://www.nytimes.com/2012/05/01/us/politics/epa-official-in-texas-resigns-over-crucify-comments.html; Kimberley A. Strassel, "The 'Crucify Them' Presidency," *Wall Street Journal*, May 4, 2012.

34. Jim Fuquay, "Resignation at EPA is Cheered and Bemoaned," *Fort Worth Star-Telegram*, May 1, 2012, http://www.star-telegram.com/living/family/moms/article3831204.html; Randy Lee Loftis, "EPA's Regional Administrator in Dallas Resigns But Says He Was Not Forced Out," *Dallas Morning News*, April 30, 2012, https://www.dallasnews.com/news/news/2012/04/30/epas-regional-administrator-in-dallas-resigns-but-says-he-was-not-forced-out; "EPA Official Right to Resign After Stir," *Austin American-Statesman*, May 3, 2012; Sharon Wilson (TXhharon), "Big Gas Mafia Wins Again, Our Beloved Dr. Al Has Resigned," *Bluedaze*, April 30, 2012, http://www.texassharon.com/2012/04/30/big-gas-mafia-wins-again-our-beloved-dr-al-has-resigned/; "Statement of Ken Kramer, Director, Lone Star Chapter, Sierra Club, on the Resignation of Dr. Al Armendariz as Region 6 EPA Administrator," Lone Star Chapter, Sierra Club, April 30, 2012; http://www.guidrynews.com/story.aspx?id=1000042859; Matthew Tresaugue, "Former EPA Watchdog Finds Sierra Club a Good Fit," *The Houston Chronicle*, August 11, 2012, http://www.chron.com/news/houston-texas/article/Former-EPA-watchdog-finds-Sierra-Club-a-good-fit-3781617.php#.
35. Kate Galbraith, "Hydraulic Fracturing Bill Could Force Disclosure," *New York Times*, March 24, 2011, http://www.nytimes.com/2011/03/25/us/25ttfracking.html; Peter Gorman, "Dry Holes," *Fort Worth Weekly*, April 20, 2011, https://www.fwweekly.com/2011/04/20/dry-holes/; Peter Gorman, "Shale Regulation: Drilled," *Fort Worth Weekly*, May 25, 2011, https://www.fwweekly.com/2011/05/25/shale-regulation-drilled/; Lowell Brown and Peggy Heinkel-Wolfe, "Session Yields Mixed Results," *Denton Record-Chronicle*, June 5, 2011, http://www.dentonrc.com/news/news/2011/06/05/session-yields-mixed-results; "Texas Becomes 1st to Require Fracking Disclosure," *Fort Worth Star-Telegram*, June 20, 2011.
36. Robert Francis, "Update: Study of Barnett Shale Quakes Cites 'Plausible Cause,'" *Fort Worth Business Press*, March 8, 2010; John-Laurent Tronche, "Some North Texans Suspect Drilling Behind Recent Quakes," *Fort Worth Business Press*, June 15, 2010; Robert Francis, "Chesapeake Disposal Well

Closure Could Spur Recycling Efforts," *Fort Worth Business Press*, August 17, 2009.

37. Jim Fuquay, "Injection Wells Seen as Possible Cause of Earthquakes," *Fort Worth Star-Telegram*, January 10, 2014, http://www.star-telegram.com/news/business/article3842249.html; Jim Fuquay, "Azle Crowd Frustrated by Lack of Answers About Quakes," *Fort Worth Star-Telegram*, January 7, 2014, http://www.star-telegram.com/news/local/article3841536.html.
38. Jack Z. Smith, "Wastewater from Natural Gas Drilling Is Made Clean," *Fort Worth Star-Telegram*, October 24, 2010.
39. Charles G. Grote and Thomas W. Grimshaw, "Fact-Based Regulation for Environmental Protection in Shale Gas Development," (Austin: The Energy Institute, The University of Texas at Austin, 2012), 3, 5, 7–9, 19–24. The experts actually included three persons from the Environmental Defense Fund.
40. Vicki Vaughn, "Report Blasts UT Hydraulic Fracturing Study," *Midland Reporter-Telegram*, December 8, 2012.

CONCLUSION

1. Max B. Baker, "Exploration on Hiatus: RigData Reports Barnett Shale Rig Count Down to One," *Midland Reporter-Telegram*, March 29, 2015.
2. The Perryman Group, *Economic and Fiscal Contribution of the Barnett Shale: Impact of Oil and Gas Exploration and Production on Business Activity and Tax Receipts in the Region and State* (Waco: The Perryman Group, 2014), 2–6, 11, 21.
3. Jack Z. Smith, "Has the Barnett Shale Left Its Best Days Behind?" *Fort Worth Star-Telegram*, November 12, 2011; Jim Fuquay, "Oil Boom in the Barnett: Natural Gas Prices Have Fizzled, but in the Northwest Shale, Another Commodity is Drawing Interest," *Fort Worth Star-Telegram*, June 24, 2012; Jim Fuquay, "Report Questions Long-Term Productivity of Gas Wells in Barnett Shale," *Fort Worth Star-Telegram*, February 13, 2013.
4. Jim Fuquay, "Oil Boom in the Barnett"; Joseph Rago, "The Oilmen to Thank at Your Next Fill-Up: The Weekend Interview with Mark Papa," *Wall Street Journal*, December 6, 2014.
5. Jim Fuquay, "Barnett 2.0 Is Hard to Define; New Owners Will Have a Growing Impact," *Fort Worth Star-Telegram*, October 14, 2012; "Pioneer Natural

Resources to Sell Barnett Shale Assets," *Oil & Gas Journal*, September 18, 2012, http://www.ogj.com/articles/2012/09/pioneer-natural-resources-to-sell-barnett-shale-assets.html.

6. Two illuminating and highly readable accounts of McClendon's creation of and career at Chesapeake may be found in Russell Gold's *The Boom: How Fracking Ignited the American Energy Revolution and Changed the World* (New York: Simon & Schuster, 2014), chapters 8 and 9, and Gary Sernovitz, *The Green and Black: The Complete Story of the Shale Revolution, the Fight Over Fracking, and the Future of Energy* (New York: St. Martin's Press, 2016), 36–39, 48, 53).
7. Gold, *The Boom*, 197–209; "Chesapeake Energy Company Fined $765,000 over Royalty Reports," PennEnergy, April 2, 2013, http://www.pennenergy.com/articles/pennenergy/2013/04/chesapeake-energy-company-fined-765000–over-royalty-reports.html; Jim Fuquay, "Land Owners Find Gas Firm Taking Royalties Bite," *Fort Worth Star-Telegram*, April 28, 2013; Jim Fuquay, "Chesapeake Sued Again over Barnett Royalty Payments," *Fort Worth Star-Telegram*, August 19, 2013; Caty Hirst, "City of Fort Worth Joins Arlington in Suit Against Chesapeake Energy," *Fort Worth Star-Telegram*, October 18, 2013; Jim Malewitz, "Fort Worth Sues Driller, Citing Millions in Lost Royalties," *New York Times*, December 21, 2013.
8. Jim Fuquay, "Chesapeake Loses Appeal of Royalty Suit Brought by Fort Worth Family," *Fort Worth Star-Telegram*, March 7, 2014; Jay F. Marks, "Chesapeake Energy Panel: No Intentional Misconduct by CEO Aubrey McClendon," The *Oklahoman*, February 21, 2013; Brianna Bailey, "Oklahoma County District Court Judge Dismisses Chesapeake Shareholder Lawsuit on Air Travel," The *Journal Record*, October 19, 2012; Gold, *The Boom*, 203, 208; Brianna Bailey, "Aubrey McClendon's New Firm in Dispute with Ohio Coal Company over Businesses' Names," The *Oklahoman*, August 8, 2013.
9. Erin Ailworth, "Chesapeake Boss Faces Tall Order Amid Bust," *Wall Street Journal*, November 27, 2015; Ryan Dezember, "Write-Downs Abound as Prices Keep Oil in Ground," *Wall Street Journal*, September 14, 2015; Associated Press, "Fort Worth to Accept $6M Royalties Deal with Chesapeake," *Midland Reporter-Telegram*, March 24, 2016; Matt Zborowski, "Chesapeake Leaves Barnett, Transfers Interest to First Reserve Unit," *Oil & Gas Journal*, August 11, 2016, http://www.ogj.com/articles/2016/08/chesapeake-leaves-barnett-transfers-interest-to-first-reserve-unit.html.

10. Bradley Olson, Ryan Dezember, and Lynn Cook, "Indicted Oil Titan Killed in Car Crash," *Wall Street Journal*, March 3, 2016; Ryan Dezember, Bradley Olson, and Erin Ailworth, "McClendon Bet Big on Comeback," *Wall Street Journal*, March 8, 2016.
11. "Range Sees 200 MMcfd from Marcellus by 2010," *Oil & Gas Journal*, December 21, 2009, 8; Stephen Bull, "Marcellus Shale Gas Play Entry Opportunities Abound," *Oil & Gas Journal*, February 1, 2010, 34–40; James Mason, "Bakken's Maximum Potential Oil Production Rate Explored," *Oil & Gas Journal*, April 2, 2012, 76–85; Paula Dittrick, "Bakken Oil Producers Turn to Railroads, Pipelines," *Oil & Gas Journal*, December 5, 2011, 43–44; Jennifer Hiller, "Crude Oil Will Continue Rolling by Train," *Midland Reporter-Telegram*, July 30, 2013.
12. Scott Stevens, Michael Godec, and Keith Moodhe, "New Plays Emerge, Although Environmental Issues Arise," *Oil & Gas Journal*, October 19, 2009, 39–40; Robert W. Gilmer, Raúl Hernandez, and Keith R. Phillips, "Oil Boom in Eagle Ford Shale Brings New Wealth to South Texas," *Southwest Economy*, Second Quarter 2012, 3–7; Ryan Holeywell, "Eagle Ford Produces Its Billionth Barrel of Oil," *Midland Reporter-Telegram*, December 7, 2014; Jennifer Hiller, "Price Plunge Puts Eagle Ford on Edge," *Midland Reporter-Telegram*, January 18, 2015; Jennifer Hiller, "Eagle Ford Oil Firms Urged to be Good Neighbors," *Midland Reporter-Telegram*, September 21, 2014.
13. Tom Fowler, "Second Life for Old Oil Field," *Wall Street Journal*, November 20, 2013; Ashley Eady, "Cline Shale Play Part of Unconventional Permian Game," *Lubbock Avalanche-Journal*, February 9, 2013; Jim Fuquay, "Once Unprofitable Oil Play is Now Booming," *Fort Worth Star-Telegram*, April 7, 2013; "Wolfcamp Shale Graduates to 'World Class' Play," *Unconventional Oil & Gas Report*, September—October 2013, 1,6; Rye Druzin, "Permian Oil Production Above 2M Barrels a Day," *Midland Reporter-Telegram*, May 12, 2015. Ironically, in the Permian Basin, the Barnett Shale did not develop as an attractive production target.
14. Rye Druzin, "API Report: Texas is Global Energy Powerhouse," *Midland Reporter-Telegram*, June 1, 2015; Conglin Xu and Laura Bell, "Global Reserves, Oil Production Show Increases for 2014," *Oil & Gas Journal*, December 1, 2014, 30–33; Emily Pickrell, "Texas Will Continue to Lead U.S. Oil Boom, Experts Say," *Midland Reporter-Telegram*, July 30, 2013; Jennifer Hiller, "American Crude Hits a 114-Year High, Led by Shale," *Midland Reporter-Telegram*, April 6, 2015; "BP: US Surpassing Saudis in Oil Output

Among World's 'Tectonic' Energy Shifts in 2014," *Oil & Gas Journal*, June 10, 2015, http://www.ojg.com/articles/2015/06/bp-us-surpassing-saudis-in-oil-output-among-world-s-tectonic-energy-shifts-in-2014.html.

15. Vicki Vaughan, "Texas Oil Jobs Hit Record, Surpass 300,000," *Midland Reporter-Telegram*, September 8, 2014; Christopher S. Rugaber, "Why Many Experts Missed This: Cheap Oil Can Hurt US Economy," *Midland Reporter-Telegram*, May 25, 2015; John Mangalonzo, "Oil Boom Brings Shortage of Truck Drivers," *Midland Reporter-Telegram*, August 4, 2013; Mary Lashley Barcella and John W. Larson, "Jobs and Income: The Impact of Shale Gas on the U.S. Economy," *Wall Street Journal*, March 8, 2012; "HIS: Unconventional Supply Chain Industries Support Growing Number of Jobs, Revenues," Oil & Gas Journal, October 6, 2014, http://www.ogj.com/articles/print/volume-112/issue-10/general-interest/his-unconventional-supply-chain-industries-support-growing-number-of-jobs-revenues.html; Paul Wiseman, "TIPRO's 'State Energy Report' Aims to be Tool for Legislators, Other Leaders," *Midland Reporter-Telegram*, April 14, 2013.
16. Rugaber, "Why Many Experts Missed This"; Mella McEwen, "Downturn Has 'Taken Bite' out of Economy," *Midland Reporter-Telegram*, July 1, 2015; Lynn Cook, "Oil Patch is Bracing for Further Cuts in Jobs," *Wall Street Journal*, July 27, 2015.
17. Paula Dittrick, "Unconventional Resource Estimates Subject to Uncertainty, Future Costs," *Oil & Gas Journal*, April 2, 2012, 36–40.
18. Russell Gold, "U.S. Energy Boom Has Room to Run," *Wall Street Journal*, September 15, 2014; Mella McEwen, "Permian Basin Daily Production to Grow by 3,000 Barrels," *Midland Reporter-Telegram*, June 10, 2015; Mella McEwen, "IHS Analysis: 'Adolescent' Wolfcamp Development Holds Tremendous Promise," *Midland Reporter-Telegram*, June 28, 2015; Rye Druzin, "Production in the Permian Up; Other Basins Wane," *Midland Reporter-Telegram*, April 7, 2015; Collin Eaton, "Occidental Pumping Savings into Permian Basin," *Midland Reporter-Telegram*, August 2, 2015; Dittrick, "Unconventional Resource Estimates," 38.
19. Arthur Berman, "Lessons from the Barnett Shale Imply Caution in other Shale Plays," *World Oil*, August 2009, 17.
20. Wayne Slater, "Energy Conference: Boom Nearly Over," *Dallas Morning News*, December 1, 2012.
21. Tom Armistead, "Bill Powers: Pickens and Stansberry Wrong, Shale Gas Production to Fall," *Pennenergy*, May 7, 2013, http://www.pennenergy

.com/articles/pennengery/2013/05/bill-powers-shale-gas-production-to-fall.html.

22. Lynn Berry, "BP Chief: Unconventional Gas 'Game Changer' in U.S.," *Midland Reporter-Telegram*, February 7, 2010; P.K. Meyer, "Shale Source Rocks a Game Changer due to 8-to-1 Resource Potential," *Oil & Gas Journal*, May 7, 2012, 72–74; Daniel Yergin, "Stepping on the Gas," *Wall Street Journal*, April 2–3, 2011; IHS Cambridge Energy Research Associates, *Fueling North America's Energy Future: The Unconventional Natural Gas Revolution and the Carbon Agenda: Executive Summary* (Cambridge: IHS CERA, 2010), ES-4; Nick Snow, "PGC Report: US Gas Resource Base Reaches a Record 2515 tcf," *Oil & Gas Journal*, September 28, 2015, 50–52.
23. Meagan S. Mauter, Vanessa R. Palmer, Yiqiao Tang, and A. Patrick Behrer, *The Next Frontier in United States Shale Gas and Tight Oil Extraction: Strategic Reduction of Environmental Impacts* (Cambridge: Energy Technology Innovation Policy Research Group, Harvard Kennedy School: Belfer Center for Science and International Affairs, March 2013), 9–10; Dittrick, "Unconventional Resources Estimates," 38.
24. John Browning, Scott W. Tinker, Svetlana Ikonnikova, Gürcam Gülen, Eric Potter, Qilong Fu, Susan Horvath, Tad Patzek, Frank Male, William Fisher, Forrest Roberts, and Ken Medlock III, "Study Develops Decline Analysis, Geologic Parameters for Reserves, Production Forecast," *Oil & Gas Journal*, August 5, 2013, 62–73; Browning *et al.*, "Barnett Study Determines Full-Field Reserves, Production Forecast," *Oil & Gas Journal*, September 2, 2013, 88–95; Jim Fuquay, "Study Says Barnett Will Still Be Producing Gas in 2050," *Fort Worth Star-Telegram*, March 1, 2013.
25. Collin Eaton, "Oil Firms Overwhelm Market, Promise New Life for Shale," *Midland Reporter-Telegram*, August 17, 2015.
26. Jeffrey Sparshott, "Oil Boom a 'Game Changer' on Trade Deficit," *Wall Street Journal*, February 6, 2015; Leon E. Panetta and Stephen J. Hadley, "The Oil-Export Ban Harms National Security," *Wall Street Journal*, May 20, 2015; Conglin Xu and Laura Bell, "Global Oil Glut Continues Despite Increasing Demand," *Oil & Gas Journal*, July 6, 2015, 26–37.
27. Amy Myers Jaffe, "How Shale Gas is Going to Rock the World," *Wall Street Journal*, May 10, 2010; Kenneth B. Medlock III, Amy Myers Jaffe, and Peter R. Hartley, "Shale Gas and U.S. National Security," (Houston: James A. Baker III Institute for Public Policy of Rice University, July 2011), 12–13.
28. Xu and Bell, "Global Oil Glut Continues," 35.

29. Chris Tomlinson, "OPEC Experiments with the Oil Industry, and the US Shale Sector is the Guinea Pig," *Midland Reporter-Telegram*, June 11, 2015; Greg Ip, "Shale Upends OPEC Bloc Party," *Wall Street Journal*, June 4, 2014; Benoit Faucon, "Oil Output Weighs on OPEC," *Wall Street Journal*, August 12, 2015; Benoit Faucon, Bill Spindle, and Summer Said, "OPEC Clout Hits New Low," *Wall Street Journal*, June 1, 2015.
30. Russell Gold, "As Rout Deepens, Oil Producers Keep on Pumping," *Wall Street Journal*, August 21, 2015.
31. Texas Department of State Health Services, *Final Report—Dish, Texas Exposure Investigation, Dish, Denton County, Texas* (Austin: Texas Department of State Health Services, May 12, 2010), 1–4.

SELECTED BIBLIOGRAPHY

DOCUMENTS FROM CITY GOVERNMENTS AND STATE AGENCIES

City of Cleburne. State of the City Report, 2010.

City of Cleburne. *Monthly Economic Development Trends Report*: July, 2010.

City of Fort Worth: Ordinances.

Railroad Commission of Texas. Office of the General Counsel. Oil and Gas Docket No. 7B-0268629: Commission Called Hearing to Consider Whether Operation of the Range Production Company Butler Unit Well No. 1H (RRC ID 253732) and the Teal Unit Well No.1H (RRC ID 253729) in the Newark East (Barnett Shale) Field, Hood County, Texas, are Causing or Contributing to Contamination of Certain Domestic Water Wells in Parker County, Texas, 2011.

Texas Commission on Environmental Quality. Ethridge, Shannon. "Health Effects Review of Ambient Air Monitoring Data Collected by Wolf Eagle Environmental Engineers and Consultants for DISH, TX." Austin: *Summary of TCEQ Actions Relating to the Barnett Shale*, 2009.

——. "Responses to Representative Lon Burnam's Oil and Gas Questions." Austin: *Summary of TCEQ Actions Relating to the Barnett Shale, Attachment 9*, 2009. http://www.oilandgaslawyerblog.com/TCEQ%2520Summary%2520with%2520Burnam%2520Q%26A%5B1%5D.pdf.

——. *Summary of TCEQ Actions Relating to the Barnett Shale, Attachment 8*, 2009. Texas Department of State Health Services. "Dish, Texas Exposure

Investigation, Dish, Denton County, Texas: Final Report." Austin: Texas Department of State Health Services, 2010.

NEWSPAPERS

Abilene Reporter-News
Austin American-Statesman
Cleburne Times-Review
Dallas Business Journal
Dallas Morning News
Dallas Observer
Decatur News
Denton Record-Chronicle
Fort Worth Business Press
Fort Worth Record
Fort Worth Star-Telegram
Fort Worth Weekly
Lubbock Avalanche-Journal
Midland Reporter-Telegram
New York Daily News
New York Examiner
New York Times
The Oklahoman
Wall Street Journal
Weatherford Democrat
Weatherford Star-Telegram
Wichita Falls Times Record News
Wise County Messenger

INDUSTRY PUBLICATIONS

AAPG Explorer
American Oil & Gas Reporter
Bulletin of the American Association of Petroleum Geologists
Drill Bit

E&P
Energy in Depth
Fort Worth Basin Oil & Gas Magazine
Oil and Gas Financial Journal
Oil and Gas Investor
Oil and Gas Journal
Oil Voice
Oil Weekly
PennEnergy
Powell Barnett Shale Newsletter
Search and Discovery
World Oil

REPORTS BY SCHOLARS AND CONSULTANTS

Armendariz, Al. "Emissions from Natural Gas Production in the Barnett Shale Area and Opportunities for Cost-Effective Improvements." Environmental Defense Fund. Austin, Texas. 2009.

Bowman, Jeff R. "Air Quality in the Barnett Shale—A Thirty-one Part Scientific Study." 2010-2011. http://www.barnettshalenews.com/documents/ntxairstudy/Air%20Quality%20in%20the%20Barnett%20Shale%20by%20Jeff%20R%20Bowman%20MPH%20March%2026%202011.pdf

Groat, Charles G., and Thomas W. Grimshaw. "Fact-Based Regulation for Environmental Protection in Shale Gas Development." Austin, Texas. The Energy Institute, University of Texas at Austin. Feb. 2012.

R. W. Harden & Associates, Inc. "Northern Trinity/Woodbine GAM Assessment of Groundwater Use in the Northern Trinity Aquifer Due to Urban Growth and Barnett Shale Development." Austin: Texas Water Development Board, 2007.

IHS Cambridge Energy Research Associates. "Fueling North America's Energy Future: The Unconventional Natural Gas Revolution and the Carbon Agenda: Executive Summary." Cambridge: IHS CERA, 2010.

Industrial Hygiene and Safety Technology, Inc. "Review of Ambient Air Monitoring Project and Related Communication Concerning Gas Well Emissions Provided for the City of Fort Worth Environmental Management Department." 2009.

Lee, Michael B. "Pipelines and Eminent Domain in Texas." Austin: Current Texas Law and Policy, Eminent Domain Institute, 2008.

Mauter, Meagan S., Vanessa R. Palmer, Yiqiao Tang, and A. Patrick Behrer. "The Next Frontier in United States Shale Gas and Tight Oil Extraction: Strategic Reduction of Environmental Impacts." Cambridge, Mass: Belfer Center for Science and International Affairs, Harvard Kennedy School, March 2013.

Medlock, Kenneth B., Amy Myers Jaffe, and Peter R. Hartley. "Shale Gas and U.S. National Security." Houston: Rice University, James A. Baker III Institute for Public Policy, July 2011.

Scott, Gayle, and J. M. Armstrong. "The Geology of Wise County, Texas." *The University of Texas Bulletin, No. 3224*. Austin: The University of Texas, 1932.

Subra, Wilma. "Evaluation of Town of DISH, Texas Ambient Air Monitoring Analysis." http://www.earthworksaction.org/library/detail/evaluation_of_town_of_dish_texas_ambient_air_monitoring_analysis.

———. "Results of Health Survey of Current and Former DISH/Clark, Texas Residents." 2009. http://www.earthworksaction.org/library/detail/health_survey_results_of_current_and_former_dish_clark_texas_residents.

Theodori, Gene. "Public Opinion Research on Urban Gas Drillers." Huntsville: Center for Rural Studies Research & Outreach, Sam Houston State University, 2009.

The Perryman Group. "An Enduring Resource: A Perspective on the Past, Present, and Future Contribution of the Barnett Shale to the Economy of Fort Worth and the Surrounding Area." Waco, Tx.: The Perryman Group, 2009.

———. "The Economic and Fiscal Contribution of the Barnett Shale: Impact of Oil and Gas Exploration and Production on Business Activity and Tax Receipts in the Region and State." Waco, Tx.: The Perryman Group, 2014.

———. "Long-Term Effect of the Barnett Shale." N.p: The Perryman Group, 2007.

Udden, J. A., and Drury McN. Phillips. *A Reconnaissance Report on the Geology of the Oil and Gas Fields of Wichita and Clay Counties, Texas*. Bureau of Economic Geology and Technology. Bulletin of the University of Texas. Number 246. Austin: University of Texas, 1912.

Wang, Zhongmin, and Alan Krupnick. "A Retrospective Review of Shale Gas Development in the United States: What Led to the Boom?" *Resources for the Future*. Washington, DC: April, 2013.

Weinstein, Bernard L., and Terry L. Clower. "The Economic and Fiscal Impacts of Devon Energy in Denton, Tarrant, and Wise Counties." Denton: University of North Texas, 2004.

——. "The Economic and Fiscal Impacts of Devon Energy Corporation in the Barnett Shale of North Texas: An Update." Denton: University of North Texas, 2006.

Winton, W.M. *The Geology of Denton County*. Bureau of Economic Geology. University of Texas Bulletin 2544. Austin: University of Texas, 1925.

Wolf Eagle Environmental Engineers and Consultants. "Deborah Rogers/ HDLA, LLC (d. b. a. Deborah's Farmstead) Ambient Air Monitoring Analysis Project: Final Report." 2009.

Wolf Eagle Environmental. "Town of DISH, Texas Ambient Air Monitoring Analysis Final Report." 2009.

BOOKS AND ARTICLES

Beebe, B. Warren, ed. *Natural Gases of North America: A Symposium in Two Volumes*: volume 2. Tulsa: American Association of Petroleum Geologists, 1968.

Burt, J. Zack. "Playing the 'Wild Card' in the High-Stakes Game of Urban Drilling: Unconscionability in the Early Barnett Shale Gas Leases." *Texas Wesleyan Law Review*. Fall, 2008: 1-30.

Burrough, Bryan. *The Big Rich: The Rise and Fall of the Greatest Texas Oil Fortunes*. New York: The Penguin Press, 2009.

Campbell, Randolph B. *Gone to Texas: A History of the Lone Star State*. New York: Oxford University Press, 2003.

Castaneda, Christopher J., and Joseph A. Pratt. *From Texas to the East: A Strategic History of Texas Eastern Corporation*. College Station: Texas A&M University Press, 1993.

Castaneda, Christopher J., and Clarence M. Smith. *Gas Pipelines and the Emergence of America's Regulatory State: A History of Panhandle Eastern Corporation, 1828-1993*. New York: Cambridge University Press, 1996.

Childs, William R. *The Texas Railroad Commission: Understanding Regulation in America to the Mid-twentieth Century*. College Station: Texas A&M University Press, 2005.

Clark, James A. *The Chronological History of the Petroleum and Natural Gas Industries*. Houston: Clark Book Co., 1963.

Clark, James A., and Michael T. Halbouty. *Spindletop*. New York: Random House, 1952.

Deffeyes, Kenneth S. *Hubbert's Peak: The Impending World Oil Shortage*. Princeton: Princeton University Press, 2009.

Ely, Glen Sample. "Gone from Texas and Trading with the Enemy: New Perspectives on Civil War West Texas." *Southwestern Historical Quarterly* 110:4 (April 2007): 438-463.

Fancher, George H., with Robert L. Whiting, and James H. Cretsinger. *The Oil Resources of Texas: A Reconnaissance Survey of Primary and Secondary Reserves of Oil*. Austin: The Texas Petroleum Research Committee, 1954.

Francaviglia, Richard V. *The Cast Iron Forest: A Natural and Cultural History of the North American Cross Timbers*. Austin: University of Texas Press, 2000.

Gardner, Frank J. *Rinehart's North Texas Oil: A Correlation of Characteristics of the Oil Fields of North and North Central Texas*. Houston: Rinehart Oil News Company of Texas, 1941.

Gilmer, Robert W., Raul Hernandez, and Keith R. Phillips. "Oil Boom in Eagle Ford Shale Brings New Wealth to South Texas." *Southwest Economy*, Second Quarter 2010, 3-7.

Gold, Russell. *The Boom: How Fracking Ignited the American Energy Revolution*. New York: Simon & Schuster, 2014.

Gonzalez, Catherine Troxell. *Rhome: A Pioneer History*. Burnet: Eakin Press, 1979.

Goodwyn, Lawrence. *Texas Oil, American Dreams: A Study of the Texas Independent Producers and Royalty Owners Association*. Austin: Texas State Historical Association, 1996.

Historical Committee of the Fort Worth Petroleum Club, *Oil Legends of Fort Worth*. N.p.: Taylor Publishing Company, 1993.

Irving, Washington. *A Tour of the Prairies*. Edited and with an Introductory Essay by John Francis McDermott. Norman: University of Oklahoma Press, 1956.

Jordan, Terry G., with John L. Bean, Jr., and William M. Holmes. *Texas: A Geography*. Boulder: Westview Press, 1984.

King, John O. *Joseph Stephen Cullinan: A Study of Leadership in the Texas Petroleum Industry: 1897-1937*. Nashville: Vanderbilt University Press, 1970.

Knight, Oliver. *Fort Worth: Outpost on the Trinity*. Fort Worth: Texas Christian University Press, 1990.

Kutchin, Joseph W. *How Mitchell Energy & Development Corp. Got Its Start and How It Grew: An Oral History and Narrative Overview*. The Woodlands: Mitchell Energy & Development Corp., 1998.

Linsley, Judith Walker, Ellen Walker Rienstra, and Jo Ann Stiles. *Giant Under the Hill: A History of the Spindletop Oil Discovery at Beaumont, Texas, in 1901.* Austin: Texas State Historical Association, 2002.

Martin, Charles A., ed. *Petroleum Geology of the Fort Worth Basin and Bend Arch Area.* Dallas: Dallas Geological Society, 1982.

Mason, Richard. "Lost Seas and Forgotten Climes: Petroleum and Geologists in North Texas." *West Texas Historical Association Year Book:* 1987, 131-152.

McDaniel, Robert W., and Henry C. Dethloff. *Pattillo Higgins and the Search for Texas Oil.* College Station: Texas A&M University Press, 1989.

Merrill, Karen R. *The Oil Crisis of 1973-1974: A Brief History with Documents.* Boston: Bedford/St. Martin's, 2007.

Olien, Diana Davids and Roger M. Olien. *Oil in Texas: The Gusher Age, 1895-1945.* Austin: University of Texas Press, 2002.

Olien, Roger M. and Diana Davids Olien. *Easy Money: Oil Promoters and Investors in the Jazz Age.* Chapel Hill: University of North Carolina Press, 1990.

——. *Oil and Ideology: The Cultural Creation of the American Petroleum Industry.* Chapel Hill: University of North Carolina Press, 2000.

——. *Oil Booms: Social Change in Five Texas Towns.* Lincoln: University of Nebraska Press, 1982.

——. *Wildcatters: Texas Independent Oilmen.* Austin: Texas Monthly Press, 1984.

Owen, Edgar Wesley. *Trek of the Oil Finders: A History of Exploration for Petroleum.* Tulsa: American Association of Petroleum Geologists, 1975.

Pate, J'Nell L. *Arsenal of Defense: Fort Worth's Military Legacy.* Denton: Texas State Historical Association, 2011.

——. *Livestock Legacy: The Fort Worth Stockyards, 1887-1987.* College Station: Texas A&M University Press, 1988.

Powers, Stephen. *Afoot and Alone: A Walk from Sea to Sea by the Southern Route, Adventures And Observations in Southern California, New Mexico, Arizona, Texas, Etc.* Edited and with an introduction by Harwood P. Hinton. Austin: The Book Club of Texas, 1995.

Rawlins, Rachael. "Planning for Fracking on the Barnett Shale: Urban Air Pollution, Improving Health Regulation, and the Role of Local Governments." *Virginia Environmental Law Journal* 31: 226-306.

Rhinehart, Marilyn D. *A Way of Work and a Way of Life: Coal Mining in Thurber, Texas, 1888-1926.* College Station: Texas A&M University Press, 1992.

Rister, Carl Coke. *Oil! Titan of the Southwest*. Norman: University of Oklahoma Press, 1949.

Sabin, Paul. *Crude Politics: The California Oil Market, 1900-1940*. Berkeley: University of California Press, 2005.

Schmelzer, Janet L. *Where the West Begins: Fort Worth and Tarrant County*. Northridge, CA.: Windsor Publications, Inc., 1985.

Sernovitz, Gary. *The Green and the Black: The Complete Story of the Shale Revolution, the Fight over Fracking, and the Future of Energy*. New York: St. Martin's Press, 2016.

Simmons, Matthew R. *Twilight in the Desert: The Coming Saudi Oil Shock and the World Economy*. New York: John Wiley & Sons, Inc., 2005.

Snider, L. C. *Oil and Gas in the Mid-Continent Fields*. Oklahoma City: Harlow Publishing Co., 1920.

Spratt, John S. *The Road to Spindletop: Economic Change in Texas, 1875-1901*. Austin: University of Texas Press, 1970.

Spratt, John S., Sr. *Thurber, Texas: The Life and Death of a Company Coal Town*, ed. Harwood P. Hinton. Abilene: State House Press, 2005.

Steward, Dan B. *The Barnett Shale Play: Phoenix of the Fort Worth Basin: A History*. Fort Worth: The Fort Worth Geological Society and the North Texas Geological Society, 2007.

Talbert, Robert H. *Cowtown-Metropolis: Case Study of a City's Growth and Structure*. Fort Worth: Texas Christian University, 1956.

Tussing, Arlon R., and Bob Tippee. *The Natural Gas Industry: Evolution, Structure, and Economics*. 2nd ed. Tulsa: Pennwell, 1995.

Tyler, Ron, et al, editors. *The New Handbook of Texas*. Austin: The Texas State Historical Association, 1996.

United States Department of the Interior. United States Geological Survey. Bulletin 629: *Natural Gas Resources of Parts of North Texas*: Eugene Wesley Shaw, "Gas in the Area North and West of Fort Worth." Washington, DC: Government Printing Office, 1916.

———. United Stsates Geological Survey. *Water Supply Paper 317*: C. H. Gordon, "Geology and Underground Waters of the Wichita Region, North Central Texas." Washington, DC: Government Printing Office, 1913.

Vietor, Richard H. K. *Energy Policy in America Since 1945: A Study in Business—Government Relations*. Cambridge: Cambridge University Press, 1987.

Warner, C. A. *Texas Oil and Gas Since 1543*. Houston: Gulf Publishing Company, 1939.

Wheeler, H. A. "Wild Boom in the North Texas Oil Fields." *Engineering and Mining Journal*. March 27, 1920, 741-747.

Woodard, Don. *Black Diamonds! Black Gold!: The Saga of Texas Pacific Coal & Oil Company*. Lubbock: Texas Tech University Press, 1998.

INTERVIEWS

Judy Alter, interviewed by Diana Davids Hinton, July 27, 2009, Fort Worth, Texas.

David H. Arrington, interviewed by Diana Davids Hinton, October 16, 2009, Midland, Texas

Larry Brogdon, interviewed by Diana Davids Hinton, July 28, 2010, Fort Worth, Texas.

Jerry Cash, interviewed by Diana Davids Hinton, July 29, 2010, Cleburne, Texas.

Ted Collins, interviewed by Diana Davids Hinton, July 23, 2009, Midland, Texas.

L. Decker Dawson, C. Roy Tobias, and Steve C. Jumper, interviewed by Diana Davids Hinton, July 6, 2009, Midland, Texas.

Nowell Donovan, interviewed by Diana Davids Hinton, June 23, 2014, by telephone.

Roger Harmon, interviewed by Diana Davids Hinton, July 29, 2010, Cleburne, Texas.

Ralph O. Harvey, interviewed by Diana Davids Olien and Roger M. Olien, July 25, 1996, Wichita Falls, Texas.

Rick Holden, interviewed by Diana Davids Hinton, March 1, 2013, Cleburne, Texas.

Ed Ireland, interviewed by Diana Davids Hinton, July 31, 2009, Fort Worth, Texas.

Dick Lowe, interviewed by Diana Davids Hinton, July 27, 2010, Fort Worth, Texas.

George P. Mitchell, interviewed by Diana Davids Hinton, July 7, 2010, Houston, Texas.

Charles Moncrief, interviewed by Diana Davids Hinton, July 27, 2010, Fort Worth, Texas.

Mike Moncrief, interviewed by Diana Davids Hinton, February 28, 2013, Fort Worth, Texas.

Ken Morgan, interviewed by Diana Davids Hinton, July 30, 2010, Fort Worth, Texas.

Marty Searcy, interviewed by Diana Davids Hinton, February 20, 2009, Fort Worth, Texas.

Dan B. Steward, interviewed by Diana Davids Hinton, March 4, 2010, Dallas, Texas.

Hollis Sullivan, interviewed by Diana Davids Hinton, July 27, 2010, Fort Worth, Texas.

David Watts, interviewed by Diana Davids Hinton, July 20, 2009, Midland, Texas.

Dan Williams, interviewed by Diana Davids Hinton, July 30, 2010, Fort Worth, Texas.

Dan and Debora Young, interviewed by Diana Davids Hinton, July 28, 2010, Fort Worth, Texas.

George M. Young, Jr., interviewed by Diana Davids Hinton, July 28, 2009, Fort Worth, Texas.

INDEX

INDEX

Diana Davids Hinton is Professor of History and holds the J. Conrad Dunagan Chair of Regional and Business History at the University of Texas of the Permian Basin. She has a PhD in history from Yale University, and, with Roger M. Olien, is coauthor of six books on the history of the American petroleum industry.